D1242872

AMERICA THROUGH EUROPEAN EYES

A YEARS RESIDENCE
IN THE UNITED STATES

A

YEARS RESIDENCE

IN THE

UNITED STATES OF AMERICA

IN THREE PARTS

BY

WILLIAM COBBETT

[1818-1819]

REPRINTS OF ECONOMIC CLASSICS

AUGUSTUS M. KELLEY · PUBLISHERS
NEW YORK 1969

First Edition 1818-1819

(London: Printed for Sherwood, Neely and Jones,
Paternoster Row, 1818-1819)

Reprinted 1969 by
Augustus M. Kelley • Publishers
New York New York 10010

SBN 6780 0516 8

Library of Congress Catalogue Card Number
70–85139

PRINTED IN THE UNITED STATES OF AMERICA
by SENTRY PRESS, NEW YORK, N. Y. 10019

A

YEAR'S RESIDENCE,

IN THE

UNITED STATES OF AMERICA.

Treating of the Face of the Country, the Climate, the Soil, the Products, the Mode of Cultivating the Land, the Prices of Land, of Labour, of Food, of Raiment; of the Expenses of House-keeping, and of the usual manner of Living; of the Manners and Customs of the People; and of the Institutions of the Country, Civil, Political, and Religious.

IN THREE PARTS.

By WILLIAM COBBETT.

PART I.

Containing,—I. A Description of the Face of the Country, the Climate, the Seasons, and the Soil, the facts being taken from the Author's daily notes during a whole year.—II. An Account of the Author's agricultural experiments in the Cultivation of the *Ruta Baga*, or Russia, or Swedish Turnip, which afford proof of what the climate and soil are.

LONDON:

PRINTED FOR SHERWOOD, NEELY, AND JONES,

PATERNOSTER-ROW.

1818.

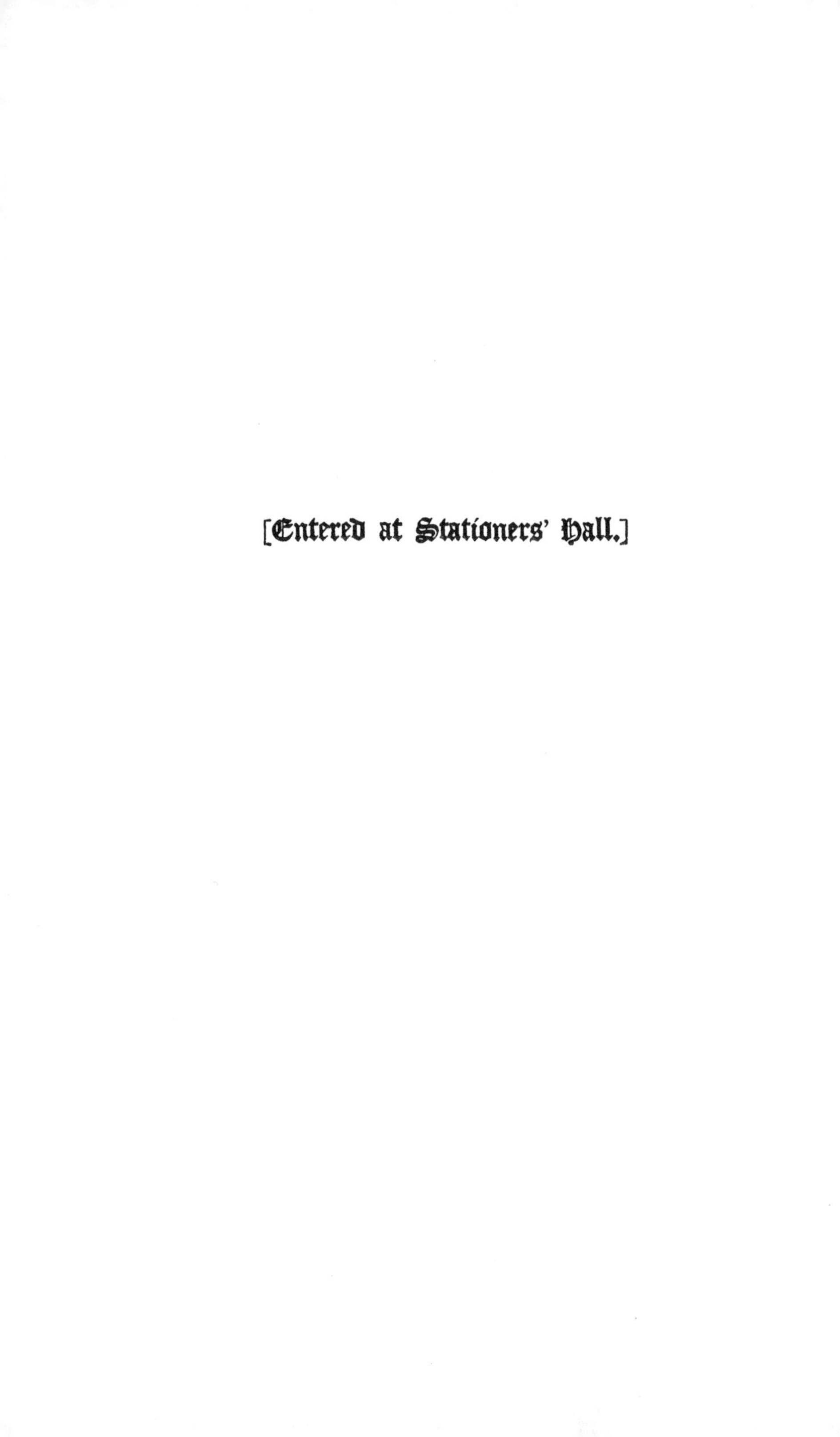

CONTENTS OF PART I.

GENERAL PREFACE

TO THE

THREE PARTS.

---

1. THROUGHOUT the whole of this work it is my intention to *number* the paragraphs, from *one* to the *end* of each PART. This renders the business of *reference* more easy than it can be rendered by any mode in my power to find out; and, easy reference saves a great deal of paper and print, and also, which ought to be more valuable, a great deal of *time*, of which an industrious man has never any to spare. To desire the reader to look at paragraph *such a number* of *such a part*, will frequently, as he will find, save him both money and labour; for, without this power of reference, the paragraph, or the substance of it, would demand being repeated in the place, where the reference would be pointed out to him.

2. Amongst all the publications, which I have yet seen, on the subject of the United States, as a country to *live in*, and especially to *farm* in, I have never yet observed one that conveyed to Englishmen any thing like a correct notion of the matter. Some writers of *Travels* in these States have jolted along in the stages from place to place, have lounged away their time with the idle part of their own countrymen, and, taking every thing different from what they left at home for the effect of ignorance, and every thing not servile to be the effect of insolence, have described the country as unfit for a civilized being to reside in. Others, coming with a resolution to find *every thing* better than at home, and weakly deeming themselves pledged to find climate, soil, and all blessed by the effects of freedom, have painted the country as a perfect paradise; they have seen nothing but blooming orchards and smiling faces.

3. The account, which I shall give, shall be that of actual *experience*. I will say what I *know* and what I have *seen* and what I have *done*. I mean to give an account of a Year's

RESIDENCE, ten months in this Island and two months in Pennsylvania, in which I went back to the first ridge of mountains. In the course of the THREE PARTS, of which this work will consist, each part making a small volume, every thing which appears to me useful to persons intending to come to this country shall be communicated; but, more especially that which may be useful to *farmers;* because, as to such matters, I have ample experience. Indeed, this is the *main thing;* for this is really and truly a *country of farmers.* Here, Governors, Legislators, Presidents, all are farmers. A farmer here is not the poor dependent wretch that a Yeoman Cavalry man is, or that a Treason-Jury man is. A farmer here depends on nobody but *himself* and on his own proper means; and, if he be not at his ease, and even rich, it must be his own fault.

4. To make men clearly see what they may do in any situation of life, one of the best modes, if not the very best, is, to give them, in detail, an account of what one has done oneself in that same situation, and how and when and where one has done it. This, as far as relates to

farming and *house-keeping* in the country, is the mode that I shall pursue. I shall give an account of what I have done; and, while this will convince any good farmer, or any man of tolerable means, that *he* may, if he will, do the same, it will give him an idea of the climate, soil, crops, &c. a thousand times more neat and correct, than could be conveyed to his mind by any general description, unaccompanied with actual experimental accounts.

5. As the expressing of this intention may, perhaps, suggest to the reader to ask, how it is that much can be known on the subject of *Farming* by a man, who, for *thirty-six* out of *fifty-two* years of his life has been a *Soldier or a Political Writer*, and who, of course, has spent so large a part of his time in garrisons and in great cities, I will beg leave to satisfy this natural curiosity before-hand.

6. Early habits and affections seldom quit us while we have vigour of mind left. I was brought up under a father, whose talk was chiefly about his garden and his fields, with regard to which he was famed for his skill and his exemplary neatness. From my very infancy,

from the age of six years, when I climbed up the side of a steep sandrock, and there scooped me out a plot four feet square to make me a garden, and the soil for which I carried up in the bosom of my little blue smock-frock (or hunting-shirt), I have never lost one particle of my passion for these healthy and rational and heart-cheering pursuits, in which every day presents something new, in which the spirits are never suffered to flag, and in which, industry, skill, and care are sure to meet with their due reward. I have never, for any eight months together, during my whole life, been without a garden. So sure are we to overcome difficulties where the heart and mind are bent on the thing to be obtained!

7. The beautiful plantation of *American Trees* round my house at Botley, the seeds of which were sent me, at my request, from Pennsylvania, in 1806, and some of which are now *nearly forty feet high*, all sown and planted by myself, will, I hope, long remain as a specimen of my perseverance in this way. During my whole life I have been *a gardener*. There is no part of the business, which, first or last, I have

not performed with my own hands. And, as to *it* I owe very little to *books*, except that of TULL; for I never read a good one in my life, except a French book, called the *Manuel du Jardinier*.

8. As to *farming*, I was bred at the plough-tail, and in the Hop-Gardens of Farnham in Surrey, my native place, and which spot, as it so happened, is the neatest in England, and, I believe, in the whole world. All there is a garden. The neat culture of the hop extends its influence to the fields round about. Hedges cut with shears and every other mark of skill and care strike the eye at Farnham, and become fainter and fainter as you go from it in every direction. I have had, besides, great experience in farming for several years of late; for, one man will gain more knowledge in a year than another will in a life. It is the *taste* for the thing that really gives the knowledge.

9. To this taste, produced in me by a desire to imitate a father whom I ardently loved, and to whose very word I listened with admiration, I owe no small part of my happiness, for a greater proportion of which very few men ever

had to be grateful to God. These pursuits, innocent in themselves, instructive in their very nature, and always tending to preserve health, have a constant, a never-failing source, of recreation to me; and, which I count amongst the greatest of their benefits and blessings, they have always, in my house, supplied the place of the card-table, the dice-box, the chess-board and the lounging bottle. Time never hangs on the hands of him, who delights in these pursuits, and who has books on the subject to read. Even when shut up within the walls of a prison, for having complained that Englishmen had been flogged in the heart of England under a guard of German Bayonets and Sabres; even then, I found in these pursuits a source of pleasure inexhaustible. To that of the whole of our English books on these matters, I then added the reading of all the valuable French books; and I then, for the first time, read that Book of all Books on husbandry, the work of JETHRO TULL, to the principles of whom I owe more than to all my other reading and all my experience, and of which principles I hope to find time to give a sketch, at least, in some future PART of this work.

10. I wish it to be observed, that, in any thing which I may say, during the course of this work, though *truth* will compel me to state facts, which will, doubtless, tend to induce farmers to leave England for America, I *advise* no one so to do. I shall set down in writing nothing but what is *strictly true*. I myself am bound to England for life. My notions of allegiance to country; my great and anxious desire to assist in the restoration of her freedom and happiness; my opinion that I possess, in some small degree, at any rate, the power to render such assistance; and, above all the other considerations, my unchangeable attachment to the people of England, and especially those who have so bravely struggled for our rights: these bind me to England; but, I shall leave others to judge and to act for themselves.

WM. COBBETT.

North Hempsted, Long Island, 21*st April,* 1818.

A

YEAR'S RESIDENCE,

&c.

CHAP. I.

Description of the Situation and Extent of Long Island, and also of the Face of the Country, and an account of the Climate, Seasons, and Soil.

LONG ISLAND is situated in what may be called the *middle* climate of that part of the United States, which, coastwise, extends from Boston to the Bay of Chesapeake. Farther to the South, the cultivation is chiefly by negroes, and farther to the North than Boston is too cold and arid to be worth much notice, though, doubtless, there are to be found in those parts good spots of land and good farmers. Boston is about 200 miles to the North of me, and the Bay of Chesapeake about the same distance to the South. In speaking of the *climate* and *seasons*, therefore, an allowance must be made, of hotter or colder, earlier or later, in a degree proportioned to those distances; because I can speak positively only of the very spot, at which I have resided. But this is a matter of very little consequence; seeing that every part has its seasons first or last. All the difference is, that, in some parts of the immense space of which I have spoken, there is a little more summer than in other parts. The same crops will, I believe, grow in them all.

12. The situation of Long Island is this: It is about 130 miles long, and, on an average, about 8 miles broad. It extends in length from the Bay of the City of New York to within a short distance of the State of Rhode Island. One side of it is against the sea, the other side looks across an arm of the sea into a part of the State of New York (to which Long Island belongs) and into a part of the State of Connecticut. At the end nearest the city of New York it is separated from the scite of that city, by a channel so narrow as to be crossed by a Steam-Boat in a few minutes; and this boat, with another near it, impelled by a team of horses, which work in the boat, form the mode of conveyance from the Island to the city, for horses, waggons, and every thing else.

13. The Island is divided into three counties, King's county, Queen's county, and the county of Suffolk. King's county takes off the end next New York city, for about 13 miles up the Island; Queen's county cuts off another slice about thirty miles further up; and all the rest is the county of Suffolk. These counties are

divided into townships. And, the municipal government of Justices of the Peace, Sheriffs, Constables, &c. is in nearly the English way, with such differences as I shall notice in the *second part* of this work.

14. There is a *ridge of hills*, which runs from one end of the Island to the other. The two sides are flats, or, rather, very easy and imperceptible slopes towards the sea. There are no rivers, or rivulets, except here and there a little run into a bottom which lets in the sea-water for a mile or two as it were to meet the springs. *Dryness* is, therefore, a great characteristic of this Island. At the place where I live, which is in Queen's county, and very nearly the middle of the Island, crosswise, we have no water, except in a well seventy feet deep, and from the clouds; yet, we never experience a want of water. A large rain-water cistern to take the run from the house, and a duck-pond to take that from the barn, afford an ample supply; and I can truly say, that as to the article of water, I never was situated to please me so well in my life before. The rains come about once in fifteen days; they come in abundance for about twenty-four hours; and then all is fair and all is dry again immediately: yet here and there, especially *on the hills*, there are *ponds*, as they call them here; but, in England, they would be called *lakes*, from their extent as well

as from their depth. These, with the various trees which surround them, are very beautiful indeed.

15. The *farms* are so many plots originally scooped out of woods; though in King's and Queen's counties the land is generally pretty much deprived of the woods, which, as in every other part of America that I have seen, are beautiful beyond all description. The Walnut of two or three sorts, the Plane, the Hickory, Chesnut, Tulip Tree, Cedar, Sassafras, Wild Cherry (sometimes 60 feet high); more than fifty sorts of Oaks; and many other trees, but especially the Flowering Locust, or Acacia, which, in my opinion, surpasses all other trees, and some of which, in this Island, are of a very great height and girt. The Orchards constitute a feature of great beauty. Every farm has its orchard, and, in general, of cherries as well as of apples and pears. Of the cultivation and crops of these, I shall speak in another Part of the work.

16. There is one great draw-back to all these beauties, namely, the *fences*; and, indeed, there is another with us South of England people; namely, the general (for there are many exceptions) slovenliness about the homesteads, and particularly about the *dwellings of labourers.* Mr. BIRKBECK complains of this; and, indeed, what a contrast with the homesteads and cot-

tages, which he left behind him near that exemplary spot, Guildford in Surrey! Both blots are, however, easily accounted for.

17. The *fences* are of *post and rail.* This arose, in the first place, from the abundance of timber that men knew not how to dispose of. It is now become an affair of *great expense* in the populous parts of the country; and, that it might, with great advantage and perfect ease, be got rid of, I shall clearly show in another part of my work.

18. The *dwellings and gardens, and little out-houses of labourers,* which form so striking a feature of beauty in England, and especially in Kent, Sussex, Surrey, and Hampshire, and which constitute a sort of fairy-land, when compared with those of the labourers in France, are what I, for my part, most feel the want of seeing upon Long Island. Instead of the neat and warm little cottage, the yard, cow-stable, pig-sty, hen-house, all in miniature, and the garden, nicely laid out and the paths bordered with flowers, while the cottage door is crowned with a garland of roses or honey-suckle; instead of these, we here see the labourer content with a shell of boards, while all around him is as barren as the sea-beach; though the natural earth would send melons, the finest in the world, creeping round his door, and though there is no English shrub, or flower, which will not

grow and flourish here. This want of attention in such cases is hereditary from the first settlers. They found land so plenty, that they treated small spots with contempt. Besides, the *example* of neatness was wanting. There were no gentlemen's gardens, kept as clean as drawing-rooms, with grass as even as a carpet. From endeavouring to imitate perfection men arrive at mediocrity; and, those who never have seen, or heard of perfection, in these matters, will naturally be slovens.

19. Yet, notwithstanding these *blots*, as I deem them, the face of the country, in summer, is very fine. From December *to May*, there is not a *speck of green*. No green-grass and turnips, and wheat, and rye, and rape, as in England. The frost comes and sweeps all vegetation and verdant existence from the face of the earth. The wheat and rye *live;* but, they lose all their verdure. Yet the state of things *in June*, is, as to crops, and fruits, much about what it is in England; for, when things do begin to grow, they grow indeed; and the general harvest for *grain* (what we call *corn*) is a full month *earlier* than in the *South* of England!

20. Having now given a sketch of the face of the country, it only remains for me to speak in this place of the *Climate* and *Seasons*, because I shall sufficiently describe the *Soil*, when I

come to treat of my own actual experience of it. I do not like, in these cases, *general descriptions*. Indeed, they must be very imperfect; and, therefore, I will just give a copy of *a Journal*, kept by myself, from the 5th of May, 1817, to the 20th of April, 1818. This, it appears to me, is the best way of proceeding; for, then, there can be no deception; and, therefore, I insert it as follows.

1817.

May 5. Landed at New York.

6. Went over to Long Island. Very fine day, warm as *May* in England. The Peach-trees going out of bloom. Plum trees in full bloom.

7. Cold, sharp, East wind, just like that which makes the old debauchees in London shiver and shake.

8. A little frost in the night, and a warm day.

9. Cold in the shade and hot in the sun.

10. The weather has been dry for some time. The grass is only beginning to grow a little.

11. Heavy thunder and rain in the night, and all this day.

12. Rain till noon. Then warm and beautiful.

1817.

May 13. Warm, fine day. Saw, in the garden, lettuces, onions, carrots, and parsnips, just come up out of the ground.

14. Sharp, drying wind. People travel with great coats, to be guarded against the morning and evening air.

15. Warm and fair. The farmers are beginning to plant their *Indian Corn*.

16. Dry wind, warm in the sun. Cherry trees begin to come out in bloom. The Oaks show no green yet. The Sassafras in flower; or whatever else it is called. It resembles the Elder flower a good deal.

17. Dry wind. Warmer than yesterday. An English April morning, that is to say, a sharp April morning, and a *June* day.

18. Warm and fine. Grass pushes on. Saw some Lucerne in a warm spot, 8 inches high.

19. Rain all day. Grass grows apace. People plant potatoes.

20. Fine and warm. A good cow sells, with a calf by her side, for 45 dollars. A steer, two years old, 20 dollars. A working ox, five years old, 40 dollars.

21. Fine and warm day; but the morn-

1817.

May 21. ing and evening coldish. The cherry-trees in full bloom, and the pear-trees nearly the same. Oats, sown in April, up, and look extremely fine.

22. Fine and warm.—Apple-trees fast coming into bloom. Oak buds breaking.

23. Fine and warm. — Things grow away. Saw kidney-beans up and looking pretty well. Saw some beets coming up. Not a sprig of parsley to be had for love or money. What improvidence! Saw some cabbage plants up and in the fourth leaf.

24. Rain at night and all day to-day. Apple-trees in full bloom, and cherry-bloom falling off.

25. Fine and warm.

26. Dry coldish wind, but hot sun. The grass has pushed on most furiously.

27. Dry wind. Spaded up a corner of ground and sowed (in the natural earth) *cucumbers* and *melons*. Just the time they tell me.

28. Warm and fair.

29. Cold wind; but the sun warm.

1817.

May 29. *No fires* in parlours now, except now-and-then in the mornings and evenings.

30. Fine and warm.—Apples have dropped their blossoms. And now the grass, the wheat, the rye, and every thing, which has stood the year, or winter through, appear to have *overtaken* their like in Old England.

31. Coldish morning and evening.

June 1. Fine warm day; but saw a man, in the evening, *covering* something in a garden. It was *kidney-beans*, and he feared a *frost!* To be sure, they are very tender things. I have had them nearly killed in England, by *June* frosts.

2. Rain and warm.—The oaks and all the trees, except the Flowering Locusts, begin to look greenish.

3. Fine and warm.—The *Indian Corn* is generally come up; but looks yellow in consequence of the cold nights and little frosts.—N. B. I ought here to describe to my English readers what this same *Indian Corn* is.——The Americans call it *Corn*, by way of eminence, and wheat, rye, barley and oats, which we confound

1817.

June 3. under the name of *corn*, they confound under the name of *grain*. The Indian Corn, in its ripe seed state, consists of an *ear*, which is in the shape of a *spruce-fir apple*. The grains, each of which is about the bulk of the largest marrow-fat pea, are placed all round the stalk, which goes up the middle, and this little stalk, to which the seeds adhere, is called the Corn *Cob*. Some of these ears (of which from 1 to 4 grow upon a plant) are more than *a foot* long; and I have seen many, each of which weighed more than *eighteen ounces*, avoirdupois weight. They are long or short, heavy or light, according to the land and the culture. I was at a Tavern, in the village of North Hempstead, last fall (of 1817) when I had just read, in the Courier, English news-paper, of a Noble Lord who had been sent on his travels to France at ten years of age, and who, from his high-blooded ignorance of vulgar things, I suppose, had *swallowed a whole ear of corn*, which, as the newspaper told us, had well nigh choaked the Noble Lord. The

1817.

June 3. Landlord had just been showing me some of his fine ears of Corn; and I took the paper out of my pocket, and read the paragraph: "What!" said he, "swallow a *whole* "*ear of corn at once!* No wonder "that they have swallowed up poor "Old John Bull's substance." After a hearty laugh, we explained to him, that it must have been *wheat* or *barley*. Then he said, and very justly, that the Lord must have been a much greater fool than a hog is.— The plant of the Indian corn grows, upon an average, to about 8 feet high, and sends forth the most beautiful leaves, resembling the broad leaf of the water-flag. It is planted in hills or rows, so that the plough can go between the standing crop. Its stalks and leaves are *the best* of fodder, if carefully stacked; and its grain is good for every thing. It is eaten by man and beast in all the various shapes of whole corn, meal, cracked, and every other way that can be imagined. It is tossed down to hogs, sheep, cattle, in the whole ear. The two former thresh for

1817.

June 3. themselves, and the latter eat *cob* and all. It is eaten, and is a very delicious thing, in its half-ripe, or *milky* state; and *these* were the "*ears of corn,*" which the Pharisees complained of the Disciples for plucking off to eat on the Sabbath Day; for, how were they to eat *wheat ears*, unless after the manner of the "Noble Lord" above mentioned? Besides, the Indian Corn is a native of Palestine. The French, who, doubtless, brought it originally from the Levant, call it *Turkish Corn*. The *Locusts*, that John the Baptist lived on, were not (as I used to wonder at when a boy) the noxious vermin that devoured the land of Egypt; but the *bean*, which comes in the long pods borne by the three-thorned *Locust-tree*, and of which I have an abundance here. The *wild honey* was the honey of wild bees; and the hollow trees here contain swarms of them. The trees are cut, sometimes, in winter, and the part containing the swarm, brought and placed near the house. I saw this lately in Pennsylvania.

1817.

June 4. Fine rain. Began about ten o'clock.

5. Rain nearly all day.

6. Fine and warm. Things grow surprizingly.

7. Fine and warm. Rather cold at night.

8. Hot.

9. Rain all day. The wood green, and so beautiful! The leaves look so fresh and delicate! But, the Flowering Locust only begins to show leaf. It will, by and by, make up, by its beauty, for its shyness at present.

10. Fine warm day. The cattle are up to their eyes in grass.

11. Fine warm day. Like the very, very finest in England in June.

12. Fine day. And, when I say fine, I mean really fine. Not a cloud in the sky.

13. Fine and hot. About as hot as the hottest of our English July weather in common years. Lucerne, 2½ feet high.

14. Fine and hot; but, we have always a *breeze* when it is hot, which I did not formerly find in Pennsylvania. This arises, I suppose, from our nearness to the sea.

1817.

June 15. Rain all day.

16. Fine, beautiful day. Never saw such fine weather. Not a morsel of *dirt*. The ground sucks up all. I walk about and work in the land in shoes made of deer-skin. They are dressed *white*, like breeches-leather. I began to *leave off my coat* to day, and do not expect to put it on again till October. My hat is a white chip, with broad brims. Never better health.

17. Fine day. The partridges (miscalled quails) begin to sit. The orchard full of birds' nests; and, amongst others, a dove is sitting on her eggs in an apple tree.

18. Fine day. Green peas fit to gather in pretty early gardens, though only of the common hotspur sort. May-duke cherries begin to be ripe.

19. Fine day. But, now comes my alarm! The *musquitoes*, and, still worse, the common *house-fly*, which used to plague us so in Pennsylvania, and which were the only things I ever disliked belonging to the climate of America. Musquitoes are bred in *stagnant water*, of which here is none.

1817.

June 19. Flies are bred in *filth*, of which none shall be near me as long as I can use a shovel and a broom. They will follow *fresh meat* and *fish*. Have neither, or be very careful. I have this day put all these precautions in practice; and, now let us see the result.

20. Fine day. Carrots and parsnips, *sown on the 3d and 4th instant*, all up, and *in rough leaf!* Onions up. The whole garden green in 18 days from the sowing.

21. Very hot. Thunder and heavy rain at night.

22. Fine day. May-duke cherries ripe.

23. Hot and close. Distant thunder.

24. Fine day.

25. Fine day. White-heart and black-heart cherries getting ripe.

26. Rain. Planted out cucumbers and melons. I find I am rather late.

27. Fine day.

28. Fine day. Gathered cherries for *drying* for winter use.

29. Fine day.

30. Rain all night. People are planting out their cabbages for the winter crop.

1817.

July 1. Fine day. Bought 20 bushels of *English* salt for *half a dollar a bushel!*

2. Fine day.

3. Fine day.

4. Fine day. Carrots, sown 3d June, 3 inches high.

5. Very hot day. *No flies yet.*

6. Fine hot day. Currants ripe. Oats in haw. Rye nearly ripe. Indian corn two feet high. Hay-making nearly done.

7. Rain and thunder early in the morning.

8. Fine hot day. Wear no waistcoat now, except in the morning and evening.

9. Fine hot day. Apples to make puddings and pies; but our housekeeper does not know how to make an apple pudding. She puts the pieces of apple-amongst the batter! She has not read Peter Pindar.

10. Fine hot day. I work in the land morning and evening, and write in the day in a north room. The *dress* is now become a very conveni-ent, or, rather, a very little incon-

1817.

July 10. venient, affair. Shoes, trowsers, shirt and hat. No plague of dressing and undressing!

11. Fine hot day in the morning, but began to grow dark in the afternoon. A sort of haze came over.

12. Very hot day. The common black cherries, the little red honey cherries, all ripe now, and falling and rotting by the thousands of pounds weight. But this place which I rent is remarkable for abundance of cherries. Some *early peas*, sown in the second week in *June*, fit for the table. This is thirty days from the time of sowing. *No flies yet! No musquitoes!*

13. Hot and heavy, like the pleading of a quarter-sessions lawyer. No *breeze* to-day, which is rarely the case.

14. Fine day. The Indian corn four feet high.

15. Fine day. We eat turnips sown on the second of June. Early cabbages (a gift) sown in May.

16. Fine hot day. Fine young onions, sown on the 8th of June.

17. Fine hot day. Harvest of wheat,

1817.

July 17. rye, oats and barley, half done. But, indeed, what is it to do when the weather does so much!

18. Fine hot day.

19. Rain all day.

20. Fine hot day, and some wind. All dry again as completely as if it had not rained for a year.

21. Fine hot day; but heavy rain at night. *Flies, a few.* Not more than in England. My son John, who has just returned from Pennsylvania, says they are as great torments there as ever. At a friend's house (a farm-house) there, *two quarts of flies* were caught in *one window in one day!* I do not believe that there are two quarts in all my premises. But, then, I cause all *wash* and *slops* to be carried forty yards from the house. I suffer no peelings or greens, or any rubbish, to lie near the house. I suffer no fresh meat to remain more than one day fresh in the house. I proscribe all fish. Do not suffer a dog to enter the house. Keep all pigs at a distance of sixty yards. And

1817.

July 21. sweep all round about once every week at least.

22. Fine hot day.

23. Fine hot day. *Sowed Buck-wheat* in a piece of very poor ground.

24. Fine hot day. Harvest (for *grain*) nearly over. The main part of the *wheat*, &c. is put into *Barns*, which are very large and commodious. Some they put into small *ricks*, or *stacks*, out in the fields, and there they stand, *without any thatching*, 'till they are wanted to be taken in during the winter, and, sometimes they remain out for a whole year. Nothing can prove more clearly than this fact, the great difference between this climate and that of England, where, as every body knows, such stacks would be mere heaps of muck by January, if they were not, long and long before that time, carried clean off the farm by the wind. The crop is sometimes *threshed* out in the field by the feet of horses, as in the South of France. It is sometimes carried into the barn's floor, where three or four

1817.

July 24. horses, or oxen, going *abreast* trample out the grain as the sheaves, or swarths are brought in. And this explains to us the humane precept of MOSES, "not to *muzzle* the ox "as he *treadeth out the grain*," which we country people in England cannot make out. I used to be puzzled, too, in the story of RUTH, to imagine how BOAZ could be busy amongst his threshers in the height of harvest.—The weather is so fine, and the grain so dry, that, when the wheat and rye are threshed by the flail, the sheaves are barely untied, laid upon the floor, receive a few raps, and are then tied up, clean threshed, for straw, without the order of the straws being in the least changed! The ears and butts retain their places in the sheaf, and the band that tied the sheaf before ties it again. The straw is as bright as burnished gold. Not a speck in it. These facts will speak volumes to an English farmer who will see with what ease work must be done in such a country.

25. Fine hot day. Early pea, men-

1817.

July 25. tioned before, *harvested,* in forty days from the sowing. *Not more flies than in England.*

26. Fine broiling day. The Indian Corn grows away now, and has, each plant, at least *a tumbler full of water standing in the sockets of its leaves,* while the sun seems as if it would actually burn one. Yet we have *a breeze;* and, under these fine shady Walnuts and Locusts and Oaks, and on the fine grass beneath, it is very pleasant. Woodcocks begin to come very thick about.

27. Fine broiler again. Some friends from England here to-day. We spent a pleasant day; drank success to the Debt, and destruction to the Boroughmongers, in gallons of milk and water.—*Not more flies than in England.*

28. Very, very hot. The Thermometer 85 degrees in the shade; but *a breeze.* Never slept better in all my life. No covering. A sheet under me, and a straw bed. And then, so happy to have no clothes to put on but shoes and trowsers! My window looks to the East. The

1817.

July 28. moment the Aurora appears, I am in the Orchard. It is impossible for any human being to lead a pleasanter life than this. How I pity those, who are *compelled* to endure the stench of cities; but, for those who remain there without being compelled, I have no pity.

29. Still the same degree of heat. I measured a water-melon runner, which grew eighteen inches in the last 48 hours. The *dews* now are equal to showers, I frequently, in the morning, wash hands and face, feet and legs, in the dews on the high grass. The Indian Corn shoots up now so beautifully!

30. Still melting hot.

31. Same weather.

August 1. Same weather. I take off two shirts a day wringing wet. I have a clothes-horse to hang them on to dry. Drink about 20 good tumblers of milk and water every day. No ailments. Head always clear. Go to bed by day-light very often. Just after the hens go to roost, and rise again with them.

2. Hotter and hotter, I think; but, in

1817.

August 2. this weather we always have our friendly breeze.—*Not a single musquito yet.*

3. Cloudy and a little shattering of rain; but not enough to lay the dust.

4. Fine hot day.

5. A very little rain. Dried up in a minute. Planted cabbages with *dust* running into the holes.

6. Fine hot day.

7. Appearances forebode rain.—I have observed that, when rain is approaching, the *stones* (which are the rock stone of the country), with which a piazza adjoining the house is paved, *get wet.* This wet appears, at first, at the top of each round stone, and, then, by degrees, goes all over it. Rain is *sure* to follow. It has never missed; and, which is very curious, the rain lasts exactly as long as the stones take to get all over wet before it comes! The stones dry again *before the rain ceases.* However, this foreknowledge of rain is of little use here; for, when it comes, it is sure to be *soon gone;* and to be succeeded by a sun, which restores all to rights.

1817.

August 8. I wondered, at first, why I never saw any *barometers* in people's houses, as almost every farmer has them in England. But, I soon found, that they would be, if perfectly true, of no use. *Early Pears ripe.*

8. Fine Rain. It comes pouring down.

9. Rain still, which has now lasted 60 hours.—Killed a lamb, and, in order to keep it fresh, sunk it down into the *well.*—The wind makes the Indian Corn bend.

10. Fine clear hot day. The grass, which was brown the day before yesterday, is already beautifully green. In one place, where there appeared no signs of vegetation, the grass is *two inches high.*

11. Heavy rains at night.

12. Hot and close.

13. Hot and close.

14. Hot and close. No breezes these three days.

15. Very hot indeed. 80 degrees in a North aspect at 9 in the evening. *Three* wet shirts to day. Obliged to put on a dry shirt *to go to bed in.*

16. Very hot indeed. 85 degrees, the thermometer hanging under the Lo-

1817.

Aug. 16. cust trees and swinging about with the breeze. The *dews* are now like heavy showers.

17. Fine hot day. Very hot. I fight the Borough-villains, stripped to my shirt, and with nothing on besides, but shoes and trowsers. Never ill; no head-aches; no muddled brains. The *milk and water* is a great cause of this. I live on salads, other garden vegetables, apple-puddings and pies, butter, cheese (*very good* from Rhode Island), eggs, and bacon. Resolved to have no more fresh meat, 'till cooler weather comes. Those who have a mind to swallow, or be swallowed by, *flies*, may eat fresh meat for me.

18. Fine and hot.

19. Very hot.

20. Very hot; but a breeze every day and night.—Buckwheat, sown 23rd July, 9 inches high, and, poor as the ground was, looks very well.

21. Fine hot day.

22. Fine hot day.

23. Fine hot day. I have now got an English woman servant, and she makes us famous apple-puddings.

1817.

Aug. 23. She says she has never read Peter Pindar's account of the dialogue between the King and the Cottage-woman; and yet she knows very well how to get the apples within side of the paste. N. B. No man ought to come here, whose wife and daughters cannot make puddings and pies.

24. Fine hot day.

25. Fine hot day.

26. Fine hot day.

27. Fine hot day. Have not seen a cloud for many days.

28. Windy and rather coldish. Put on cotton stockings and a waistcoat with sleeves. Do not like this weather.

29. Same weather. Do not like it.

30. Fine and hot again. Give a great many apples to hogs. Get some hazle-nuts in the wild grounds. Larger than the English: and much about the same taste.

31. Fine hot day. Prodigious *dews*.

Sept. 1. Fine and hot.

2. Fine and hot.

3. Famously hot. Fine breezes. Began imitating the Disciples, at least

1817.

Sept. 3. in their *diet;* for, to day, we began "*plucking the ears of corn*" in a patch planted in the garden on the second of June. But, we, in imitation of Pindar's pilgrim, take the liberty to *boil* our Corn. We shall not starve now.

4. Fine and hot. 83 degrees under the Locust-trees.

5. Very hot indeed, but fair, with our old breeze.

6. Same weather.

7. Same weather.

8. Same weather.

9. Rather hotter. We, amongst seven of us, eat about 25 ears of Corn a day. With *me* it wholly supplies the place of bread. It is the choicest gift of God to man, in the way of food. I remember, that ARTHUR YOUNG observes, that the proof of *a good climate* is, that Indian Corn come to perfection in it. Our Corn is very fine. I believe, that a wine-glass full of *milk* might be squeezed out of one ear. No wonder the Disciples were tempted to pluck it when they were hungry, though it was on the Sabbath day!

1817.

Sept. 10. Appearances for rain; and, it is time; for my neighbours begin to cry out, and our rain-water cistern begins to shrink. The *well* is there, to be sure; but, to pull up water from 70 feet is no joke, while it requires nearly as much sweat to get it up, as we get water.

11. No rain; but cloudy. 83 degrees in the shade.

12. Rain and very hot in the morning. Thunder and heavy rain at night.

13. Cloudy and cool. Only 55 degrees in shade.

14. Cloudy and cool.

15. Fair and cool. *Made a fire* to write by. Don't like this weather.

16. Rain, warm.

17. Beautiful day. Not very hot. Just like a fine day in July in England after a rain.

18. Same weather. Wear stockings now and a waistcoat and neck-handkerchief.

19. Same weather. Finished our Indian Corn, which, on less than 4 rods, or perches, of ground, produced 447 ears. It was singularly well cultivated. It was the long yellow Corn.

1817.

Sept. 19. Seed given me by my excellent neighbour, Mr. John Tredwell.

20. Same weather.

21. Same weather.

22. Same weather.

23. Cloudy and hotter.

24. Fine rain all last night and until ten o'clock to-day.

25. Beautiful day.

26. Same weather. 70 degrees in shade. Hot as the hot days in August in England.

27. Rain all last night.

28. Very fine and warm. Left off the stockings again.

29. Very fine, 70 degrees in shade.

30. Same weather.

October 1. Same weather. Fresh meat keeps pretty well now.

2. Very fine; but, there was a little *frost* this morning, which did not, however, affect the late sown *Kidney Beans*, which are as tender as the cucumber plant.

3. Cloudy and warm.

4. Very fine and warm, 70 degrees in shade. The apples are very fine. We are now cutting them and quinces to *dry* for winter use. My

1817.

October 4. neighbours give me quinces. We are also cutting up and drying peaches.

5. Very fine and warm. Dwarf Kidney beans very fine.

6. Very fine and warm. *Cutting Buck-wheat.*

7. Very fine and warm. 65 degrees in shade at 7 o'clock this morning.—Windy in the afternoon. The wind is knocking down the *fall-pipins* for us. One picked up to-day weighed 12¼ ounces avoirdupois weight. The average weight is about 9 ounces, or, perhaps, 10 ounces. This is the finest of all apples. Hardly any *core*. Some *none at all.* The richness of the pine-apple without the roughness. If the King could have seen one of *these* in a dumpling! This is not the *Newtown Pipin,* which is sent to England in such quantities. That is a *winter* apple. Very fine at Christmas; but far inferior to this fall-pipin, taking them both in their state of perfection. It is useless to send *the trees* to England, unless the heat of the sun and the rains

Oct. 7. and the dews could be sent along with the trees.

8. Very fine, 68 in shade.

9. Same weather.

10. Same weather, 59 degrees in shade. A little white frost this morning. It just touched the lips of the kidney bean leaves; but, not those of the cucumbers or melons, which are near fences.

11. Beautiful day. 61 degrees in shade. Have not put on a *coat* yet. Wear thin stockings, or socks, waistcoat with sleeves, and neckcloth. In New York Market, ***Kidney Beans*** and *Green peas*.

12. Beautiful day. 70 degrees in shade.

13. Same weather.

14. Rain. 50 degrees in shade. Like a fine, warm, *June* rain in England.

15. Beautiful day. 56 degrees in shade. Here is *a month of October!*

16. Same weather. 51 degrees in shade.

17. Same weather, but a little warmer in the day. A *smart frost* this morning. The kidney beans, cucumber and melon plants, pretty much cut by it.

18. A little rain in the night. A most

1817.

Oct. 18. beautiful day. 54 degrees in shade. A June day for England.

19. A very *white frost* this morning. Kidney beans, cucumbers, melons, all demolished; but a beautiful day. 56 degrees in shade.

20. Another frost, and just such another day.—*Threshing Buckwheat in field.*

21. No frost. 58 degrees in shade.

22. Finest of English June days. 67 degrees in shade.

23. Beautiful day. 70 degrees in shade. Very few summers in England that have a day hotter than this. It is this fine sun that makes the fine apples!

24. Same weather precisely. Finished Buckwheat threshing and winnowing. The men have been away at a horse-race; so that it has laid out in the field, partly threshed and partly not, for five days. If rain had come, it would have been of no consequence. All would have been dry again directly afterwards. What a stew a man would be in, in England, if he had his grain lying about out of doors in this way!

1817.

Oct. 24. The *cost* of threshing and winnowing 60 bushels was 7 dollars, 1*l.* 11*s.* 6*d.* English money, that is to say, 4*s. a quarter*, or eight Winchester bushels. But, then, the *carting* was next to nothing. Therefore, though the labourers had *a dollar a day each*, the expense, upon the whole, was not so great as it would have been in England. So much does the climate do!

25. Rain. A warm rain, like a fine June rain in England. 57 degrees in shade. The late frosts have killed, or, at least, pinched the leaves of the trees; and they are now red, yellow, russet, brown, or of a dying green. Never was any thing so beautiful as the bright sun, shining through these fine lofty trees upon the gay verdure beneath.

26. Rain. Warm. 58 degrees in shade. This is the general *Indian Corn harvest.*

27. Rain. Warm. 58 degrees in shade. Put on *coat, black hat* and *black shoes.*

28. Fine day. 56 degrees in shade. Pulled up a *Radish* that weighed

1817.

Oct. 28. 12 pounds! I say *twelve*, and measured 2 feet 5 inches round. From common English seed.

29. Very fine indeed.

30. Very fine and warm.

31. Very fine. 54 degrees in shade. Gathered our last lot of winter apples.

Nov. 1. Rain all the last night and all this day.

2. Rain still. 54 degrees in shade. Warm. Things grow well. The *grass very fine* and luxuriant.

3. Very fine indeed. 56 in shade. Were it not for the colour of the leaves of the trees, all would look like June in England.

4. Very, very fine. Never saw such pleasant weather. *Digging Potatoes.*

5. Same weather precisely.

6. A little cloudy, but warm.

7. Most beautiful weather! 63 degrees in shade. N. B. This is November.

8. A little cloudy at night fall. 68 degrees in shade; that is to say, English *Summer heat* all but 7 degrees.

9. Very fine.

1817.
Nov. 10. Very fine.

11. Very fine. When I got up this morning, I found the thermometer hanging on the Locust trees, dripping with dew, at 62 degrees. *Left off my coat again.*

12. Same weather. 69 degrees in shade.

13. Beautiful day, but cooler.

14. Same weather. 50 degrees in shade. The high ways and paths as clean as a boarded floor; that is to say, from *dirt* or *mud.*

15. Gentle rain. 53 in shade. Like a gentle rain in May in England.

16. Gentle rain. Warm. 56 in shade. What a *November* for an Englishman to see! My white turnips have grown almost the whole of their growth in this month. The Swedish, planted late, grow surprisingly now, and have a luxuriancy of appearance exceeding any thing of the kind I ever saw. We have fine loaved lettuces; endive, young onions, young radishes, cauliflowers with heads five inches over. The rye fields grow beautifully. They have been *food* for *cattle* for a month, or six weeks, past.

1817.

Nov. 17. Cloudy. Warm.

18. Same weather. 55 degrees in shade.

19. *Frost*, and the ground pretty hard.

20. Very fine indeed. Warm. 55 degrees in shade.

21. Same weather.

22. Cold, damp air, and cloudy.

23. Smart frost at night.

24. } 25. } 26. } 27. } Same. Warm in the day time.

28. } 29. } Same; but more warm in the day.

30. Fine warm and beautiful day; no frost at night. 57 degrees in shade.

Dec. 1. Same weather precisely; but, we begin to fear the setting-in of winter, and I am very busy in covering up cabbages, mangle wurzle, turnips, beets, carrots, parsnips, parsley, &c. the mode of doing which (not less *useful* in England than here, though not so indispensably necessary) shall be described when I come to speak of the management of these several plants.

2. Fine warm rain. 56 in shade.

1817.

Dec. 3. 4. 5. 6. 7. 8. Very fair and pleasant, but frost sufficiently hard to put a stop to our getting up and stacking turnips. Still, however, the cattle and sheep do pretty well upon the grass, which is long and dead. Fatting oxen we feed with the greens of Ruta Baga, with some corn (Indian, mind) tossed down to them in the ear. Sheep (ewes that had lambs in spring) we kill *very fat* from the grass. No dirt. What a clean and convenient soil!

9. Thaw. No rain. We get on with our work again.

10. Open mild weather.

11. Same weather. Very pleasant.

12. Rain began last night.

13. Rain all day.

14. Rain all day. The old Indian remark is, that the winter does not *set in* till the *ponds* be *full*. It is coming, then.

15. Rain till 2 o'clock. We kill mutton now. Ewes, brought from Connecticut, and sold to me here at 2 dollars each in July, just after shearing. I sell them now alive at 3 dollars

1817.

Dec. 15. each from the grass. Killed and sent to market, they leave me the loose fat for candles, and fetch about 3 dollars and a quarter besides.

16. Sharp *North West* wind. This is the *cold American Wind.* "A *North Wester*" means all that can be imagined of *clear in summer and cold in winter.* I remember hearing from that venerable and excellent man, Mr. BARON MASERES, a very elegant eulogium on the Summer North Wester in England. This is the only public servant that I ever heard of, who refused a *proffer'd augmentation of salary!*

17. A hardish frost.

18. Open weather again.

19. Fine mild day; but began freezing at night-fall.

20. Hard frost.

21. Very sharp indeed. Thermometer down to 10 degrees; that is to say, 22 degrees colder than barely freezing.

22. Same weather. Makes us *run*, where we used to walk in the fall, and to saunter in the summer. It is no new thing to *me;* but it makes

1817.

Dec. 22. our other English people shrug up their shoulders.

23. Frost greatly abated. Stones show for *wet*. It will come, in spite of all the fine serene sky, which we now see.

24. A thaw.—Servants made a lot of *candles* from mutton and beef fat, reserving the coarser parts to make *soap*.

25. Rain. Had some English friends. Sirloin of own beef. Spent the evening in light of *own candles*, as handsome as I ever saw, and, I think, the very best I ever saw. The reason is, that the tallow is *fresh*, and that it is unmixed with *grease*, which, and staleness, is the cause, I believe, of candles *running*, and plaguing us while we are using them. What an injury is it to the farmers in England, that they dare not, in this way, use their own produce! Is it not a *mockery* to call a man *free*, who no more dares turn out his tallow into candles for his own use, than he dares rob upon the highway? Yet, it is only by means of tyranny and extortion like

1817.

Dec. 25. this, that the hellish system of funding and of Seat-selling can be upheld.

26. Fine warm day. 52 degrees in shade.

27. Cold, but little frost.

28. Same weather. Fair and pleasant. The late sharp frost has changed to a *complete yellow* every leaf of some Swedish Turnips (Ruta Baga), *left to take their chance.* It is a poor chance, I believe!

29. Same weather.

30. Rain all day.

31. Mild and clear. No frost.

1818.

Jan. 1. Same weather.

2. Same weather.

3. Heavy rain.

4. A frost that makes us jump and skip about like larks. Very seasonable for a sluggish fellow. Prepared for winter. Patched up a boarded building, which was formerly a coach-house; but, which is not so necessary to me, in that capacity, as in that of a *fowl-house.* The neighbours tell me, that the poultry will roost out on the trees

1818.

Jan. 4. all the winter, which, the weather being so *dry* in winter, is very likely; and, indeed, they *must*, if they have *no house*, which is almost universally the case. However, I mean to give the poor things *a choice*. I have *lined* the said coach-house with *corn stalks* and *leaves of trees*, and have tacked up cedar-boughs to hold the lining to the boards, and have laid a *bed of leaves* a foot thick all over the floor. I have secured all against dogs, and have made ladders for the fowls to go in at holes six feet from the ground. I have made pig-styes, lined round with cedar-boughs and well covered. A sheep-yard, for a score of ewes to have lambs in spring, surrounded with a hedge of cedar-boughs, and with a shed for the ewes to lie under, if they like. The oxen and cows are tied up in a stall. The dogs have a place, well covered, and lined with corn-stalks and leaves. And now, I can, without anxiety, sit by the fire, or lie in bed, and hear the North-Wester whistle.

5. Frost. Like what we call "*a hard* frost" in England.

1818.

Jan. 6. Such another frost at night, but a thaw in the middle of the day.

7. Little frost. Fine warm day. The sun seems loth to quit us.

8. Same weather.

9. A harder frost, and snow at night. The *fowls*, which have been peeping at my ladders for two or three evenings, and partially roosting in their house, made their *general entry* this evening! They are the best judges of what is best for them. The *turkeys* boldly set the weather at defiance, and still roost on *the top*, the ridge, of the roof, of the house. Their feathers prevent their legs from being frozen, and so it is with all poultry; but, still, a house *must*, one would think, be better than the open air at this season.

10. Snow, but *sloppy*. I am now at New York on my way to Pennsylvania. N.B. This journey into Pennsylvania had, for its principal object, an appeal to the justice of the Legislature of that State for redress for great loss and injury sustained by me, nearly twenty years ago, in consequence of the tyranny of one

1818.

Jan. 10. McKean, who was then the Chief Justice of that State. The appeal has not *yet* been successful; but, as I confidently expect, that it finally will, I shall not, at present, say any thing more on the subject.—My journey was productive of much and various observation, and, I trust, of useful knowledge. But, in this place, I shall do little more than give an account of the *weather;* reserving for the Second Part, accounts of *prices* of land, &c. which will there come under their proper heads.

11. Frost but not hard. Now at New York.

12. Very sharp frost. Set off for Philadelphia. Broke down on the road in New Jersey.

13. Very hard frost still. Found the Delaware, which divides New Jersey from Pennsylvania, frozen over. Good roads now. Arrive at Philadelphia in the evening.

14. Same weather.

15. Same weather. The question eagerly put to me by every one in Philadelphia, is: "Don't you think

1818.

Jan. 15. the city *greatly improved?*" They seem to me to confound *augmentation* with *improvement*. It always was a fine city, since I first knew it; and it is very greatly augmented. It has, I believe, nearly doubled its extent and number of houses since the year 1799. But, after being, for so long a time, familiar with London, every other place appears little. After *living* within a few hundreds of yards of Westminster Hall and the Abbey Church and the Bridge, and looking from my own windows into St. James's Park, all other buildings and spots appear mean and insignificant. I went to day to see the house I formerly occupied. How small! It is always thus: the words *large* and *small* are carried about with us in our minds, and we forget real *dimensions*. The idea, *such as it was received*, remains during our absence from the object. When I returned to England, in 1800, after an absence from the country parts of it, of sixteen years, the trees, the hedges, even the parks and woods, seemed

1818.

Jan. 15. so *small!* It made me laugh to hear little gutters, that I could jump over, called *Rivers!* The Thames was but a "*Creek!*" But, when, in about a month after my arrival in London, I went to *Farnham*, the place of my birth, what was my surprise! Every thing was become so pitifully *small!* I had to cross, in my post-chaise, the long and dreary heath of Bagshot. Then, at the end of it, to mount a hill, called Hungry Hill; and from that hill I knew that I should look down into the beautiful and fertile vale of Farnham. My heart fluttered with impatience, mixed with a sort of fear, to see all the scenes of my childhood; for I had learnt before, the death of my father and mother. There is a hill, not far from the town, called *Crooksbury Hill*, which rises up out of a flat, in the form of a *cone*, and is planted with Scotch fir trees. Here I used to take the eggs and young ones of crows and magpies. This hill was a famous object in the neighbourhood. It served as the superlative degree of

1818.

Jan. 15. height. "*As high as Crooksbury Hill*" meant, with us, the utmost degree of height. Therefore, the first object that my eyes sought was this hill. *I could not believe my eyes!* Literally speaking, I for a moment, thought the famous hill removed, and a little heap put in its stead; for I had seen, in New Brunswick, a single rock, or hill of solid rock, ten times as big, and four or five times as high! The post-boy, going down hill and not a bad road, whisked me, in a few minutes to the Bush Inn, from the garden of which I could see the prodigious *sand hill*, where I had begun my gardening works. What a *nothing!* But now came rushing into my mind, all at once, my pretty little garden, my little blue smock-frock, my little nailed shoes, my pretty pigeons that I used to feed out of my hands, the last kind words and tears of my gentle and tender-hearted and affectionate mother! I hastened back into the room. If I had looked a moment longer, I should have dropped. When I came to reflect,

1818.

Jan. 15. *what a change!* I looked down at my dress. What a change! What scenes I had gone through! How altered my state! I had dined the day before at a secretary of state's in company with Mr. *Pitt*, and had been waited upon by men in gaudy liveries! I had had nobody to assist me in the world. No teachers of any sort. Nobody to shelter me from the consequence of bad, and no one to counsel me to good, behaviour. I felt proud. The distinctions of rank, birth, and wealth, all became nothing in my eyes; and from that moment (less than a month after my arrival in England) I resolved never to bend before them.

16. Same weather. Went to see my old quaker-friends at Bustleton, and particularly my beloved friend James Paul, who is very ill.

17. Returned to Philadelphia. Little frost and a little snow.

18.–21. Moderate frost. Fine clear sky. The Philadelphians are *cleanly*, a quality which they owe chiefly to the Quakers. But, after being long and recently familiar with the towns

1818.

Jan. 21. in Surrey and Hampshire, and especially with Guildford, Alton, and Southampton, no other towns appear clean and neat, not even Bath or Salisbury, which last is much about upon a par, in point of cleanliness, with Philadelphia; and, Salisbury is deemed a very cleanly place. Blandford and Dorchester are clean; but, I have never yet seen any thing like the towns in Surrey and Hampshire. If a Frenchman, born and bred, could be taken up and carried blindfold to Guildford, I wonder what his sensations would be, when he came to have the use of his sight! Every thing near Guildford seems to have received an influence from the town. Hedges, gates, stiles, gardens, houses inside and out, and the dresses of the people. The market day at Guildford is a perfect *show* of cleanliness. Not even a carter without a clean smock-frock and closely-shaven and clean-washed face. Well may Mr. Birkbeck, who came from this very spot, think the people *dirty* in the western country! I'll engage he

1818.

Jan. 21. finds more dirt upon the necks and faces of one family of his present neighbours, than he left behind him upon the skins of all the people in the three parishes of Guildford. However, he would not have found this to be the case in Pennsylvania, and especially in those parts where the Quakers abound; and, I am told, that, in the New England States, the people are as cleanly and as neat as they are in England. The sweetest flowers, when they become putrid, stink the most; and, a nasty woman is the nastiest thing in nature.

22. Hard frost. My *business* in Pennsylvania is with the legislature. It is sitting at *Harrisburgh.* Set off to-day by stage. Fine country; fine barns; fine farms. Must speak particularly of these in another place. Got to Lancaster. The largest *in-land* town in the United States. A very clean and good town. No beggarly houses. All looks like *ease* and *plenty.*

23. Harder frost, but not very severe. Almost as cold as the weather was

1818.

Jan. 23. during the six weeks continuance of the snow, in 1814, in England.

24. The same weather continues.

25. A sort of half-thaw. Sun warm. HARRISBURGH is a new town, close on the left bank of the river SUSQUEHANNAH, which is not frozen over, but has large quantities of ice floating on its waters. All vegetation, and all appearance of green, gone away.

26. Mild weather. Hardly any frost.

27. Thaws. Warm. Tired to death of the *tavern* at HARRISBURGH, though a very good one. The cloth spread three times a day. Fish, fowl, meat, cakes, eggs, sausages; all sorts of things in abundance. Board, lodging, *civil* but not *servile* waiting on, beer, tea, coffee, chocolate. Price, a dollar and a quarter a day. Here we meet altogether: senators, judges, lawyers, tradesmen, farmers and all. I am weary of the everlasting loads of meat. *Weary of being idle.* How few such days have I spent in my whole life!

28. Thaw and rain. My business not coming on, I went to a *country*

1818.

Jan. 28. tavern, hoping there to get a room to myself, in which to read my English papers, and sit down to writing. I am now at *M'Allister's tavern*, situated at the foot of the first ridge of mountains; or rather, upon a little nook of land, close to the river, where the river has found a way through a break in the chain of mountains. Great enjoyment here. Sit and read and write. My mind is again in England. Mrs. M'ALLISTER just suits me. Does not pester me with questions. Does not cram me with meat. Lets me eat and drink what I like, and when I like, and gives mugs of nice milk. I find, here, a very agreeable and instructive occasional companion, in Mr. M'ALLISTER the elder. But, of the various useful information, that I received from him, I must speak in the second part of this work.

29. Very hard frost this morning. Change very sudden. All about the house a glare of ice.

30. Not so hard. *Icicles* on the trees on the neighbouring mountains like

1818.

Jan. 30. so many millions of sparkling stones, when the sun shines, which is all the day.

31. Same weather. Two farmers of Lycoming county had heard that William Cobbett was here. They modestly introduced themselves. What a contrast with the "*yeomanry cavalry?*"

Feb. 1. Same weather. About the same as a "*hard frost*" in England.

2. Same weather.

3. *Snow.*

4. Little snow. Not much frost. This day, thirty-three years ago, I enlisted as a soldier. I always keep the day in recollection.

5. Having been to Harrisburgh on the second, returned to M'Allister's to-day *in a sleigh.* The River begins to be frozen over. It is about a mile wide.

6. Little snow again, and hardish frost.

7. Now and then a little snow.—Talk with some *hop-growers.* Prodigious crops in this neighbourhood; but, of them in the Second Part. What would a *Farnham* man think of *thirty hundred weight* of hops upon

1818.

Feb. 7. *four hundred* hills, *ploughed* between, and the ground vines *fed off by sheep!* This is a very curious and interesting matter.

8. *A real Frost.*

9. Sharper. They say, that the thermometer is down *to* 10 *degrees below nought.*

10. A little milder; but very cold indeed. The River completely frozen over, and sleighs and foot-passengers crossing in all directions.

11. Went back again to Harrisburgh. Mild frost.

12. Not being able to bear the idea of *dancing attendance*, came to *Lancaster*, in order to see more of this pretty town. A very fine Tavern (Slaymaker's); room to myself; excellent accommodations. Warm fires. Good and clean beds. *Civil* but *not servile*, landlord. The eating still more overdone than at Harrisburgh. Never saw such profusion. I have made a bargain with the landlord: he is to give me a dish of chocolate a day, *instead of dinner.*—Frost, but mild.

13. Rain.—A real rain, but rather cold.

1818.

Feb. 14. A complete day of rain.

15. A hard frost; much about like a hard frost in the naked parts of Wiltshire.—Mr. HULME joined me on his way to Philadelphia from the city of Washington.

16. A hard frost.—Lancaster is a pretty place. No *fine* buildings; but no *mean* ones. Nothing *splendid* and nothing *beggarly*. The people of this town seem to have had the prayer of HAGAR granted them: "Give me, O Lord, neither *poverty* "nor *riches*." Here are none of those poor, wretched habitations, which sicken the sight at the *out-skirts* of cities and towns in England; those abodes of the poor creatures, who have been reduced to beggary by the cruel extortions of the rich and powerful. And, this remark applies to *all* the towns of America that I have ever seen. This is a fine part of America. *Big Barns*, and modest dwelling houses. Barns of *stone*, a *hundred feet* long and *forty wide*, with two floors, and raised roads to go into them, so that the waggons go into the

1818.

Feb. 16. *first floor up-stairs.* Below are stables, stalls, pens, and all sorts of conveniences. Up-stairs are rooms for threshed corn and grain; for tackle, for meal, for all sorts of things. In the front (South) of the barn is the cattle yard. These are very fine buildings. And, then, all about them looks so comfortable, and gives such manifest proofs of ease, plenty and happiness! Such is the country of WILLIAM PENN's settling! It is a curious thing to observe the *farm-houses* in this country. They consist, almost without exception, of a considerably large and a very neat house, with sash windows, and of a *small house*, which seems to have been *tacked on* to the large one; and, the proportion they bear to each other, in point of dimensions, is, as nearly as possible, the proportion of size between a *Cow* and *her Calf*, the latter a month old. But, as to the *cause*, the process has been the opposite of this instance of the works of nature, for, it is *the large house which has grown out of the small one.* The

1818.

Feb. 16. father, or grandfather, while he was toiling for his children, lived in the small house, constructed chiefly by himself, and consisting of rude materials. The means, accumulated in the small house, enabled a son to rear the large one; and, though, when *pride* enters the door, the small house is sometimes demolished, few sons in America have the folly or want of feeling to commit such acts of filial ingratitude, and of real self-abasement. For, what inheritance so valuable and so honourable can a son enjoy as the proofs of his father's industry and virtue? The progress of wealth and ease and enjoyment, evinced by this regular increase of the size of the farmers' dwellings, is a spectacle, at once pleasing, in a very high degree, in itself; and, in the same degree, it speaks the praise of the system of government, under which it has taken place. What a contrast with the farm-houses in England! There the *little* farm-houses are falling into ruins, or, are actually become cattle-sheds, or, at best, *cottages*,

1818.

Feb. 16. as they are called, to contain a miserable labourer, who ought to have been a farmer, as his grandfather was. Five or six farms are there *now* levelled into one, in defiance of *the law;* for, there is a law to prevent it. The *farmer* has, indeed, a *fine house;* but, what a life do his labourers lead! The cause of this sad change is to be found in the crushing taxes; and the cause of them, in the Borough usurpation, which has robbed the people of their best right, and, indeed, without which right, they can enjoy no other. They talk of the *augmented population* of England; and, when it suits the purposes of the tyrants, they boast of this *fact,* as they are pleased to call it, as a proof of the fostering nature of their government; though, just now, they are preaching up the vile and foolish doctrine of PARSON MALTHUS, who thinks, that there are *too many* people, and that they ought (those who *labour*, at least) to be *restrained from breeding so fast.* But, as to the fact, I do not believe it. There

1818.

Feb. 16. can be nothing in the shape of *proof*; for no actual enumeration was ever taken till the year 1800 We know well, that London, Manchester, Birmingham, Bath, Portsmouth, Plymouth, and all Lancashire and Yorkshire, and some other counties, have got a vast increase of miserable beings huddled together. But, look at Devonshire, Somersetshire, Dorsetshire, Wiltshire, Hampshire, and other counties. You will there see *hundreds of thousands* of acres of land, where the old marks of the plough are visible, but which have not been cultivated for, perhaps, half a century. You will, there see places, that were once considerable towns and villages, now having, within their ancient limits, nothing but a few cottages, the *Parsonage* and a *single Farm-house.* It is a curious and a melancholy sight, where an ancient church, with its lofty spire or tower, the church sufficient to contain a thousand or two or three thousand of people conveniently, now stands surrounded by a score or half a score of miserable

1818.

Feb. 16. mud-houses, with floors of earth, and covered with thatch; and this sight strikes your eye in all parts of the five Western counties of England. Surely these churches were not built without the existence of a population somewhat proportionate to their size! Certainly not; for the churches are of various sizes, and, we sometimes see them very small indeed. Let any man look at the *sides of the hills* in these counties, and also in Hampshire, where *downs*, or open lands, prevail. He will there see, not only that those hills were formerly cultivated; but, that *banks*, from distance to distance, were made by the *spade*, in order to form little flats for the plough to go, without tumbling the earth down the hill; so that the side of a hill looks, in some sort, like *the steps of a stairs*. Was this done *without hands*, and without *mouths* to consume the grain raised on the sides of these hills? The Funding and Manufacturing and Commercial and Taxing System has, by drawing wealth into great masses, drawn

1818.

Feb. 16. men also into great masses. London, the manufacturing places, Bath, and other places of dissipation, have, indeed, wonderfully increased in population. Country seats, Parks. Pleasure-gardens, have, in like degree, increased in number and extent. And, in just the same proportion has been the increase of Poor-houses, Mad-houses, and Jails. But, *the people of England*, such as FORTESCUE described them, have been *swept away* by the ruthless hand of the Aristocracy, who, making their approaches by slow degrees, have, at last, got into their grasp the substance of the whole country.

17. Frost, not very hard. Went back to Harrisburgh.

18. Same weather. Very fine. Warm in the middle of the day.

19. Same weather. — Quitted Harrisburg, very much *displeased;* but, on this subject, I shall, if possible, keep silence, till *next year*, and until *the People* of Pennsylvania have had time to reflect; to clearly understand my affair; and when they *do understand it*, I am not all afraid

1818.

Feb. 19. of receiving *justice* at their hands, whether I am *present or absent.* Slept at Lancaster. One night more in this very excellent Tavern.

20. Frost still. Arrived at Philadelphia along with my friend HULME. They are *roasting an ox on the Delaware.* The fooleries of England are copied here, and every where in this country, with wonderful avidity; and, I wish I could say, that some of the vices of our "*higher orders,*" as they have the impudence to call themselves, were not also imitated. However, I look principally at the mass of farmers; the sensible and happy farmers of America.

21. *Thaw and Rain.*—The *severe* weather is over for this year.

22. Thaw and Rain. A solid day of rain.

23. Little frost at night. Fine market. Fine meat of all sorts. As *fat mutton* as I ever saw. How mistaken Mr. Birkbeck is about American mutton!

24. Same weather. Very fair days now.

1818.

Feb. 25. Went to Bustleton with my old friend, Mr. John Morgan.

26. Returned to Philadelphia. Roads very dirty and heavy.

27. *Complete thaw;* but it will be long before the frost be out of the ground.

28. Same weather. *Very warm.* I hate this weather. Hot upon my back and melting ice under my feet. The people (those who have been lazy) are chopping away with axes the ice, which has grown out of the snows and rains, before their doors, during the winter. The hogs (best of scavengers) are very busy in the streets seeking out the bones and bits of meat, which have been flung out and frozen down amidst water and snow, during the two foregoing months. I mean including the present month. At New York (and, I think, at Philadelphia also) they have *corporation* laws to prevent hogs from being in the streets. For *what reason*, I know not, except putrid meat be pleasant to the smell of the inhabitants. But, Corporations are seldom the wisest of law-

1818.

Feb. 28. makers. It is argued, that, if there were no hogs in the streets, people would not throw out their orts of flesh and vegetables. Indeed! What would they do with those orts, then? Make their hired servants eat them? The very proposition would leave them to cook and wash for themselves. Where, then, are they to fling these effects of superabundance? Just before I left New York for Philadelphia, I saw a sow very comfortably dining upon a full quarter part of what appeared to have been a *fine leg of mutton*. How many a family in England would, if within reach, have seized this meat from the sow! And, are the tyrants, who have brought my industrious countrymen to that horrid state of misery, *never* to be called to account? Are they *always* to carry it as they now do? Every object almost, that strikes my view, sends my mind and heart back to England. In viewing the ease and happiness of this people, the contrast fills my soul with indignation, and makes it more and more the object of my life to assist

1818.

Feb. 28. in the destruction of the diabolical usurpation, which has trampled on king as well as people.

March 1. Rain. Dined with my old friend SEVERNE, an honest Norfolk man, who used to carry his milk about the streets, when I first knew him, but, who is now a man of considerable property, and, like a wise man, lives in the same modest house where he formerly lived. Excellent roast beef and plum pudding. At his house I found an Englishman, and, from *Botley* too! I had been told of such a man being in Philadelphia, and that the man said, that he had *heard* of me, "*heard of such a gentleman, but did not know much of him.*" This was odd! I was desirous of *seeing* this man. Mr. SEVERNE got him to his house. His name is VERE. I knew him the moment I saw him; and, I wondered *why* it was that he *knew so little of me.* I found, that he *wanted work,* and that he had been *assisted* by some society in Philadelphia. He said he was *lame,* and he might be a little, perhaps. *I offered him*

1818.

March 1. *work at once.* No: he wanted to have the *care* of a farm! " Go," said I, " for shame, and ask some " farmers *for work.* You will find " it *immediately,* and with good " wages. What should the people " in this country see in your face to " induce them to keep you in idle- " ness? They did not send for you. " You are a young man, and you " come from a country of able " labourers. You may be rich if " you will work. This gentleman " who is now about to cram you " with roast beef and plum pudding " came to this city nearly as poor " as you are; and, I first came to " this country in no better plight. " Work, and I wish you well; be " idle, and you ought to starve.' He told me, then, that he was a *hoop-maker ;* and yet, observe, he wanted to have the *care* of a farm!

N. B. If this book should ever reach the hands of Mr. RICHARD HUIXMAN, my excellent good friend of Chilling, I beg him to show this note to Mr. NICHOLAS FREEMANTLE of Botley. *He* will know well all about this

1818.

March 1. VERE. Tell Mr. FREEMANTLE, that the Spaniels are beautiful, that Woodcocks breed here in abundance; and tell him, above all, that I frequently think of him as a pattern of industry in business, of skill and perseverance and good humour as a sportsman, and of honesty and kindness as a neighbour. Indeed, I have pleasure in thinking of all my Botley neighbours, except the Parson, who for their sakes, I wish, however, was my neighbour *now;* for *here* he might pursue his calling very *quietly.*

2. Open weather. Went to Bustleton, after having seen Messrs. STEVENS and PENDRILL, and advised them to forward to me affidavits of what they knew about OLIVER, the spy of the Boroughmongers.

3. Frost in the morning. Thaw in the day.

4. Same weather in the night. Rain all day.

5. Hard frost. Snow 3 inches deep.

6. Hard frost. About as cold as a hard frost in January in England.

7. Same weather.

8. Thaw. Dry and fine.

1818.

March 9. Same weather. Took leave, I fear for ever, of my old and kind friend, JAMES PAUL. His brother and son promise to come and see me here. I have pledged myself to *transplant* 10 acres of Indian Corn; and, if I write, in August, and say that *it is good*, THOMAS PAUL has promised that he will come; for, he thinks that the scheme is a mad one.

10. Same weather.—Mr. VAREE, a son-in-law of Mr. JAMES PAUL, brought me yesterday to another son-in-law's, Mr. EZRA TOWNSHEND at BIBERY. Here I am amongst the thick of the Quakers, whose houses and families pleased me so much formerly, and which pleasure is all now revived. Here all is ease, plenty, and *cheerfulness.* These people are never *giggling* and never in *low-spirits.* Their minds, like their dress, are simple and strong. Their kindness is shown more in acts than in words. Let others say what they will, I have uniformly found those whom I have intimately known of this sect, sincere and upright men; and, I

1818.

March 10. verily believe, that all those charges of hypocrisy and craft, that we hear against Quakers, arise from a feeling of *envy;* envy inspired by seeing them possessed of such abundance of all those things, which are the fair fruits of care, industry, economy, sobriety, and order, and which are justly forbidden to the drunkard, the glutton, the prodigal, and the lazy. As the day of my coming to Mr. TOWNSHEND's had been announced beforehand, several of the young men, who were babies when I used to be there formerly, came to see "BILLY COBBETT," of whom they had heard and read so much. When I saw them and heard them, "*What a contrast,*" said I to myself, "with the senseless, gaudy, "up-start, hectoring, insolent, and "cruel Yeomanry Cavalry in Eng-"land, who, while they grind their "labourers into the revolt of star-"vation, gallantly sally forth with "their sabres, to chop them down "at the command of a Secretary of "State; and, who, the next mo-"ment, creep and fawn like spaniels

1818.

March 10. "before their Boroughmonger Land-"lords!" At Mr. TOWNSHEND'S I saw a man, in his service, lately from YORKSHIRE, but an Irishman by birth. He wished to have an opportunity to see me. He had read many of my "little books." I shook him by the hand, told him he had now got a good house over his head and a kind employer, and advised him *not to move for one year*, and to save his wages during that year.

11. Same open weather.—I am now at *Trenton*, in New Jersey, waiting for something to carry me on towards New York.—Yesterday, Mr. TOWNSHEND sent me on, under an escort of Quakers, to Mr. ANTHONY TAYLOR'S. He was formerly a merchant in Philadelphia, and now lives in his very pretty country-house on a very beautiful farm. He has some as fine and fat oxen as we generally see at Smithfield market in London. I think they will weigh *sixty seore each*. Fine farm yard. Every thing belonging to the *farm* good, but, what a neglectful *gardener!* Saw some *white thorns* here (brought from Eng-

1818.

March 11. land, which, if I had wanted any proof, would have clearly proved to me, that they would, with *less care*, make as good hedges here as they do at Farnham in Surrey. But, in another PART, I shall give full information upon this head. Here my escort quitted me; but, luckily, Mr. NEWBOLD, who lives about ten miles nearer Trenton than Mr. Taylor does, brought me on to his house. He is a much better gardener, or, rather, to speak the truth, has *succeeded a better*, whose example he has followed in part. But, his farm yard and buildings! This was a sight indeed! Forty head of horn-cattle in a yard, enclosed with a stone wall; and five hundred merino ewes, besides young lambs, in the finest, most spacious, best contrived, and most substantially built sheds I ever saw. The barn surpassed all that I had seen before. His house (large, commodious, and handsome) stands about two hundred yards from the turnpike road, leading from Philadelphia to New York, looks on and over the Dela-

1818.

March 11. ware which runs parallel with the road, and has, surrounding it, and at the back of it, five hundred acres of land, level as a lawn, and two feet deep in loom, that never requires a water furrow. This was the finest sight that I ever saw as to farm-buildings and land. I forgot to observe, that I saw in Mr. TAYLOR's service, another man, recently arrived from England. A Yorkshire man. He, too, wished to see me. He had got some of my "*little books*," which he had preserved, and brought out with him. Mr. TAYLOR was much pleased with him. An active, smart man; and, if he follow my advice, to remain *a year* under one roof, and save his wages, he will, in a few years, be a rich man. These men must be brutes indeed not to be sensible of the great kindness and gentleness and liberality, with which they are treated. Mr. TAYLOR came, this morning, to Mr. NEWBOLD's, and brought me on to TRENTON. I am at the stage-tavern, where I have just dined upon cold ham, cold veal, butter and cheese, and a peach-pye;

1818.
March 11. nice clean room, well furnished, waiter clean and attentive, plenty of milk; and charge *a quarter of a dollar!* I thought, that Mrs. JOSLIN at Princestown (as I went on to Philadelphia), Mrs. BENLER at Harrisburgh, Mr. SLAYMAKER at Lancaster, and Mrs. M'ALLISTER, were low enough in all conscience; but, really, this charge of Mrs. ANDERSON beats all. I had not the face to pay the waiter a quarter of a dollar; but gave him half a dollar, and told him to keep the change. He is a black man. He thanked me. But, they never *ask* for any thing. But, my vehicle is come, and now I bid adieu to Trenton, which I should have liked better, if I had not seen so many young fellows lounging about the streets, and leaning against door-posts, with quids of tobacco in their mouths, or segars stuck between their lips, and with dirty hands and faces. Mr. Birkbeck's complaint, on this score, is perfectly just.

Brunswick, New Jersey. Here I am, after a ride of about 30 miles,

1818.

March 11. since two o'clock, in what is called a Jersey-waggon, through such *mud* as I never saw before. Up to the stock of the wheel; and yet a pair of very little horses have dragged us through it in the space of *five hours.* The best horses and driver, and the worst roads I ever set my eyes on. This part of Jersey is a sad spectacle, after leaving the brightest of all the bright parts of Pennsylvania. My driver, who is a tavern-keeper himself, would have been a very pleasant companion, if he had not drunk so much spirits on the road. This is the *great misfortune* of America! As we were going up a hill very slowly, I could perceive him looking very hard at my cheek for some time. At last, he said: "I am wondering, Sir, to "see you look so *fresh* and so *young*, "considering what you have gone "through in the world;" though I cannot imagine *how* he had learnt who I was. "I'll tell you," said I, "how I have contrived the thing. "I rise early, go to bed early, eat "sparingly, never drink any thing

1818.

March 11. "stronger than small beer, shave "once a day, and wash my hands "and face clean three times a day, "at the very least." He said, that was *too much* to think of doing.

12. Warm and fair. Like an English *first of May* in point of warmth. I got to ElizabethTown Point through beds of mud. Twenty minutes too late for the steam-boat. Have to wait here at the tavern till to-morrow. Great mortification. Supped with a Connecticut farmer, who was taking on his daughter to Little York in Pennsylvania. The rest of his family he took on in the fall. He has *migrated.* His reasons were these: he has *five sons*, the eldest 19 years of age, and several daughters. Connecticut is thickly settled. He has not the means to buy farms for the sons there. He, therefore, goes and gets cheap land in Pennsylvania; his sons will assist him to clear it; and, thus, they will have a farm each. To a man in such circumstances, and "born with an "axe in one hand, and a gun in the "other," the western countries are

1818.

March 12. desirable; but not to *English farmers*, who have great skill in fine cultivation, and who can purchase near New York or Philadelphia. This YANKEY (the inhabitants of Connecticut, Rhode Island, Massachusetts and New Hampshire, only, are called *Yankeys*) was about the age of SIR FRANCIS BURDETT, and, if he had been dressed in the usual clothes of Sir Francis, would have passed for him. Features, hair, eyes, height, make, manner, look, hasty utterance at times, musical voice, frank deportment, pleasant smile. All the very fac-simile of him. I had some early York cabbage seed and some cauliflower seed in my pocket, which had been sent me from London, in a letter, and which had reached me at Harrisburgh. I could not help giving him a little of each.

13. Same weather. A fine open day. Rather a cold May-day for England. Came to New York by the steam-boat. Over to this island by another, took a little light waggon, that *whisked* me home over roads as

1818.

March 13. dry and as smooth as gravel walks in an English bishop's garden in the month of July. Great contrast with the bottomless muds of New Jersey! As I came along, saw those fields of rye, which were so green in December, now *white*. Not a single sprig of green on the face of the earth. Found that my man had *ploughed ten acres of ground*. The frost not quite clean out of the ground. It has penetrated *two feet eight inches*. The weather here has been nearly about the same as in Pennsylvania; only *less snow*, and *less rain*.

14. Open weather. Very fine. Not quite so warm.

15. Same weather. *Young chickens*. I hear of no other in the neighbourhood. This is the effect of my *warm fowl-house!* The house has been supplied with eggs *all the winter*, without any interruption. I am told, that this has been the case at no other house hereabouts. We have *now* an abundance of eggs. More than a large family can consume. We send some to market. The

1818.

March 15. fowls, I find, have wanted no feeding except during the snow, or, in the very, very cold days, they *did not come out of their house all the day.* A certain proof that they like the warmth.

16. Little frost in the morning. Very fine day.

17. Precisely same weather.

18. Same weather.

19. Same weather.

20. Same weather. Opened several pits, in which I had preserved all sorts of garden plants and roots, and apples. Valuable experiments. As useful in England as here, though not so absolutely necessary. I shall communicate these in another part of my work, under the head of *gardening*.

21. Same weather. The day like a fine May-day in England. I am writing without fire, and in my waistcoat without coat.

22. Rain all last night, and all this day.

23. Mild and fine. A sow had a litter of pigs *in the leaves under the trees.* Judge of the weather by this. The

1818.

March 23. wind blows cold; but, she has drawn together great heaps of leaves, and protects her young ones with surprising sagacity and exemplary care and fondness.

24. Same weather.

25. Still mild and fair.

26. Very cold wind. We try to get the sow and pigs into the buildings. But the pigs do not follow, and we cannot, with all our temptations of corn and all our caresses, get the sow to move without them by her side. She must remain 'till they choose to travel. How does nature, through the conduct of this animal, reproach those mothers, who cast off their new-born infants to depend on a hireling's breast! Let every young man, before he marry, read, upon this subject, the pretty poem of Mr. Roscoe, called "the Nurse;" and, let him also read, on the same subject, the eloquent, beautiful, and soul-affecting passage, in Rousseau's "*Emile.*"

27. Fine warm day. Then high wind, rain, snow, and hard frost before morning.

1818.

March 28. Hard frost. Snow 3 inches deep.

29. Frost in the night; but, all thawed in the day, and very warm.

30. Frost in night. Fine warm day.

31. Fine warm day.—As the *winter* is now gone, let us take a look back at its *inconveniences* compared with those of an *English Winter.*—We have had *three* months of it; for, if we had a few days sharp in December, we have had many very fine and *without fire* in March. In England winter really begins in November, and does not end 'till Mid-March. Here we have *greater cold;* there four times as much *wet.* I have had my great coat on only *twice,* except when sitting in *a stage,* travelling. I have had *gloves* on no oftener; for, I do not, like the Clerks of the Houses of Boroughmongers, *write in gloves.* I seldom meet a waggoner with gloves or great coat on. It is generally *so dry.* This is the great friend of man and beast. Last summer *I wrote home for nails to nail my shoes for winter.* I could find none here. What a foolish people, not to have shoe-nails! I

1818.

March 31. forgot, that it was likely, that the absence of shoe-nails argued *an absence of the want of them.* The nails are not come; and I have not wanted them. There is *no dirt,* except for about *ten days at the breaking up of the frost.* The dress of a labourer does not cost *half* so much as in England. This *dryness* is singularly favourable to all animals. They are hurt far less by *dry cold,* than by *warm drip, drip, drip,* as it is in England.—There has been nothing *green* in the *garden,* that is to say, *above ground,* since December; but, we have had, all winter, and have now, *white cabbages, green savoys, parsnips, carrots, beets, young onions, radishes, white turnips, Swedish turnips,* and *potatoes;* and all these in abundance (except *radishes,* which were a few *to try),* and always *at hand* at a minute's warning. The modes of preserving will be given in another part of the work. What can any body want *more* than these things in the garden way? However

1818.

March 31. it would be very easy to add to the catalogue. Apples, quinces, cherries, currants, peaches, *dried in the summer*, and excellent for tarts and pies. Apples in their raw state, as many as we please. My own stock being gone, I have *trucked* turnips for apples; and shall thus have them, if I please, 'till apples come again on the trees. I give two bushels and a half of Swedish turnips for one of apples; and, mind, this is on the *last day of March*.—I have here stated *facts*, whereby to judge of the winter; and I leave the English reader to judge for himself, I myself decidedly preferring the American winter.

April 1. Very fine and warm.

2. Same weather.

3. Same weather.

4. Rain all day.

5. Rain all day. Our cistern and pool full.

6. Warm, but no sun.—Turkeys begin to lay.

7. Same weather. My first spring operations in gardening are now go-

1818.

April 7. ing on; but I must reserve an account of them for another Part of my work.

8. Warm and fair.

9. Rain and rather cold.

10. Fair but cold. It rained but yesterday, and we are to-day, feeding sheep and lambs with *grain of corn*, and with *oats*, upon *the ground* in the orchard. Judge, then, of the cleanness and convenience of this soil!

11. Fine and warm.

12. Warm and fair.

13. Warm and fair.

14. Drying wind and miserably cold. Fires again in day-time, which I have not had for some days past.

15. Warm, like a fine May-day in England. We are planting out selected roots for seed.

16. Rain all last night.—Warm. Very fine indeed.

17. Fine warm day. Heavy thunder and rain at night. The *Martins* (not swallows) *are come* into the barn and are looking out scites for the habitations of their future young ones.

1818.

April 18. Cold and raw. Damp, too, which is extremely rare. The worst day I have yet seen during the year. Stops the grass, stops the swelling of the buds. The young chickens hardly peep out from under the wings of the hens. The lambs don't play, but stand *knit up*. The pigs growl and squeak; and the birds are gone away to the woods again.

19. Same weather with an Easterly wind. Just such a wind as that, which, in March, brushes round the corners of the streets of London, and makes the old, muffled-up debauchees hurry home with aching joints. Some hail to day.

20. Same weather. Just the weather to give drunkards the "blue devils."

21. Frost this morning. Ice as thick as a dollar.—Snow three times. Once to cover the ground. Went off again directly.

22. Frost and ice in the morning. A very fine day, but not warm. Dandelions blow.

23. Sharp white frost in morning. Warm and fine day.

1818.

April 24. Warm night, warm and fair day. And *here I close my Journal;* for, I am in haste to get my manuscript away; and there now wants only *ten days* to complete the year.—I resume, now, the *Numbering* of my *Paragraphs,* having begun my Journal at the close of PARAGRAPH No. 20.

21. Let us, now, take a survey, or rather glance, at the face, which nature now wears. The grass begins to afford a good deal for sheep and for my grazing English pigs, and the cows and oxen get a little food from it. The pears, apples, and other fruit trees, have not made much progress in the swelling or bursting of their buds. The buds of the weeping-willow have *bursted* (for, in spite of that conceited ass, Mr. JAMES PERRY, *to burst* is a *regular verb,* and vulgar pedants only make it irregular), and those of a *Lilac,* in a warm place, are almost *bursted,* which is a great deal better than to say, " almost *burst.*" Oh, the coxcomb! As if an absolute pedagogue like him could injure me by his *criticisms!* And, as if an error like this, even if it had been one, could have any thing to do with my capacity for developing principles, and for simplifying things, which, in their nature, are of great complexity!—The oaks, which, in England, have now their sap in *full flow,* are

here quite unmoved as yet. In the gardens in general there is *nothing green*, while, in England, they have *broccoli* to eat, early cabbages planted out, coleworts to eat, peas four or five inches high. Yet, we shall have *green peas* and *loaved cabbages as soon as they will.* We have *sprouts* from the cabbage stems preserved under cover; the Swedish turnip is giving me *greens* from bulbs planted out in March; and I have some *broccoli too*, just coming on for use. ***How*** I have got this broccoli I must explain in my *Gardener's Guide;* for write one I must. I never can leave this country without an attempt to make every farmer a gardener.—In the meat way, we have beef, mutton, bacon, fowls, a calf to kill in a fortnight's time, sucking pigs when we choose, lamb nearly fit to kill; and all of our own breeding, or our own feeding. We kill an ox, send three quarters and the hide to market and keep one quarter. Then a sheep, which we use in the same way. The bacon is always ready. Some fowls always fatting. Young ducks are just coming out to meet the green peas. Chickens (the earliest) as big as American Partridges (misnamed quails), and ready for the asparagus, which is just coming out of the ground. Eggs at all times more than we can consume. And, if there be any one, who wants *better* fare than this, let the grumbling glutton come to that poverty, which Solomon has said shall be his lot. And, the *great thing of all*, is, that here, *every man*, even every labourer, may live as well as this, if he will be *sober* and *industrious*.

22. There are *two things*, which I have not yet mentioned, and which are almost wholly wanting here, while they are so amply enjoyed in England. The *singing birds* and the *flowers*. Here are many birds in summer, and some of very beautiful plumage. There are some wild flowers, and some English flowers in the best gardens. But, generally speaking, they are birds without song, and flowers without smell. The *linnet* (more than a thousand of which I have heard warbling upon one scrubbed oak on the sand hills in Surrey), the *sky-lark*, the *goldfinch*, the *wood-lark*, the *nightingale*, the *bull-finch*, the *black-bird*, the *thrush*, and all the rest of the singing tribe are wanting in these beautiful woods and orchards of garlands. When these latter have dropped their bloom, all is gone in the flowery way. No *shepherd's rose*, no *honey-suckle*, none of that endless variety of beauties that decorate the hedges and the meadows in England. No *daisies*, no *primroses*, no *cowslips*, no *blue-bells*, no *daffodils*, which, as if it were not enough for them to charm the sight and the smell, must have names, too, to delight the ear. All these are wanting in America. Here are, indeed, birds, which bear the *name* of

robin, blackbird, thrush, and goldfinch; but, alas! the thing at Westminster has, in like manner, the *name* of parliament, and speaks the voice of the people, whom it pretends to represent, in much about the same degree that the black-bird here speaks the voice of its namesake in England.

23. *Of health*, I have not yet spoken, and, though it will be a subject of remark in another part of my work, it is a matter of too deep interest to be wholly passed over here. In the first place, as to *myself*, I have always had excellent health; but, during a year, in England, I used to have a *cold* or two; a trifling sore throat; or something in that way. *Here*, I have neither, though I was more than two months of the winter travelling about, and sleeping in different beds. My family have been more healthy than in England, though, indeed, there has seldom been any serious illness in it. We have had but *one visit from any Doctor*. Thus much, for the present, on this subject. I said, in the second Register I sent home, that this climate was *not so good as that of England*. Experience, observation, a careful attention to real facts, have convinced me that it is, *upon the whole*, a better climate; though I tremble lest the tools of the Boroughmongers should cite this as a new and most flagrant instance of *inconsistency*. England is my country, and to Eng-

land I shall return. I like it best, and shall always like it best; but, then, in the word *England*, many things are included besides climate and soil and seasons, and eating and drinking.

24. In the *Second Part* of this work, which will follow the First Part in the course of two months, I shall take particular pains to detail all that is within my knowledge, which I think likely to be useful to persons who intend coming to this country from England. I shall take every particular of the expense of supporting a family, and show what are the means to be obtained for that purpose, and how they are to be obtained. My intending to return to England ought to *deter* no one from coming hither; because, I was resolved, if I had life, to return, and I expressed that resolution before I came away. But, if there are good and virtuous men, who can do no good there, and who, by coming hither, can withdraw the fruits of their honest labour from the grasp of the Borough tyrants, I am bound, if I speak of this country at all, to tell them the real truth; and this, as far as I have gone, I have now done.

CHAP. II.

RUTA BAGA.

CULTURE, MODE OF PRESERVING, AND USES OF THE RUTA BAGA, *sometimes* CALLED THE RUSSIA, AND SOMETIMES THE SWEDISH, TURNIP.

Description of the Plant.

25. IT is my intention, as notified in the public papers, to put into print an account of all the experiments, which I have made, and shall make, in Farming and in Gardening upon this Island. I several years ago, long before tyranny showed its present horrid front in England, formed the design of sending out, to be published in this country, a treatise on the cultivation of the root and green crops, as cattle, sheep, and hog food. This design was suggested by the reading of the following passage in Mr. CHANCELLOR LIVINGSTON'S *Essay on Sheep*, which I received in 1812. After having stated the most proper means to be employed in order to keep sheep and lambs, during the winter months, he adds: "Having brought our "flocks through the winter, we come now to the

" most critical season, that is, the latter end of
" March and the month of April. At this time
" the ground being bare, the sheep will refuse
" to eat their hay, while the scanty picking of
" grass, and its purgative quality, will disable
" them from taking the nourishment that is
" necessary to keep them up. If they fall away
" their wool will be injured, and the growth of
" their lambs will be stopped, and even many
" of the old sheep will be carried off by the
" dysentery. *To provide food for this season is*
" *very difficult.* *Turnips* and *Cabbages* will
" *rot,* and bran they will not eat, after having
" been fed on it all the winter. *Potatoes,* how-
" ever, and the *Swedish turnip,* called *Ruta*
" *Baga,* may be usefully applied at this time,
" and so, I think, might *Parsnips* and *Carrots.*
" But, as few of us are in the habit of cultiva-
" ting these plants to the extent which is neces-
" sary for the support of a large flock, we
" must *seek resources more within our reach.*"
And then the Chancellor proceeds to recommend the leaving the second growth of clover *uncut,* in order to produce early shoots from sheltered buds for the sheep to eat until the coming of the natural grass and the general pasturage.

26. I was much surprised at reading this passage; having observed, when I lived in Penn-

sylvania, how prodigiously the root crops of every kind flourished and succeeded with only common skill and care; and, in 1815, having by that time had many crops of Ruta Baga exceeding *thirty tons*, or, about *one thousand five hundred heaped bushels to the acre*, at Botley, I formed the design of sending out to America a treatise on the culture and uses of that root, which, I was perfectly well convinced, could be raised with more ease here than in England, and, that it might be easily preserved during the whole year, if necessary, I had proved in many cases.

27. If Mr. CHANCELLOR LIVINGSTON, whose public-spirit is manifested fully in his excellent little work, which he modestly calls an *Essay*, could see my ewes and lambs, and hogs and cattle, at this "*critical season*" (I write on the 27th of March), with more Ruta Baga at their command than they have mouths to employ on it; if he could see me, who am on a poor exhausted piece of land, and who found it covered with weeds and brambles in the month of June last, who found no manure, and who have brought none; if he could see me overstocked, not with mouths, but with food, owing to a little care in the cultivation of this invaluable root, he would, I am sure, have reason to be convinced, that, if any farmer in the United

States is in want of food at this pinching season of the year, the fault is neither in the soil nor in the climate.

28. It is, therefore, of my mode of cultivating this root on this Island that I mean, at present, to treat; to which matter I shall add, in another PART of my work, an account of my experiments as to the MANGEL WURZEL, or SCARCITY ROOT; though, as will be seen, I deem that root, except in particular cases, of very inferior importance. The parsnip, the carrot, the cabbage, are all excellent in their kind and in their uses; but, as to these, I have not yet made, upon a scale sufficiently large here, such experiments as would warrant me in speaking with any degree of confidence. Of these, and other matters, I propose to treat in a future PART, which I shall, probably, publish towards the latter end of this present year.

29. The *Ruta Baga* is a sort of turnip well known in the State of New York, where, under the name of *Russia* turnip, it is used for the Table from February to July. But, as it may be more of a stranger in other parts of the country, it seems necessary to give it enough of description to enable every reader to distinguish it from every other sort of turnip.

30. The leaf of every other sort of turnip is of a *yellowish* green, while the leaf of the Ruta Baga is of a *bluish* green, like the green

of peas, when of nearly their full size, or like the green of a young and thrifty early Yorkshire cabbage. Hence it is, I suppose, that some persons have called it the *Cabbage-turnip*. But the characteristics the most decidedly distinctive are these: that the outside of the *bulb* of the Ruta Ruga is of a greenish hue, mixed, towards the top, with a colour bordering on a red; and, that the inside of the bulb, if the sort be true and pure, is of a *deep yellow*, nearly as deep as *that of gold*.

Mode of saving and of preserving the Seed.

31. This is rather a nice business, and should be, by no means, executed in a negligent manner. For, on the well attending to this, much of the seed depends: and, it is quite surprizing how great losses are, in the end, frequently sustained by the saving, in this part of the business, of an hour's labour or attention. I, one year, lost more than half of what would have been an immense crop, by a mere piece of negligence in my bailiff as to the seed; and I caused a similar loss to a gentleman in Berkshire, who had his seed from the same parcel that mine was taken, and who had sent many miles for it, in order to have the *best in the world*.

32. The Ruta Baga is apt to *degenerate*, if the seed be not saved with care. We, in

England, *select* the plants to be saved for seed. We examine well to find out those that run least into *neck* and *green*. We reject all such as approach at all towards a *whitish* colour, or which are even of a *greenish* colour *towards the neck*, where there ought to be a little *reddish cast*.

33. Having selected the plants with great care, we take them up out of the place where they have grown, and plant them in a plot distant from every thing of the turnip or cabbage kind which is to bear seed. In this Island, I am now, at this time, planting mine for seed (27th March), taking all our English precautions. It is probable, that they would do very well, if taken out of *a heap* to be transplanted, if well selected; but, lest this should not do well, I have kept my selected plants all the winter in the ground in my garden, well covered with corn-stalks and leaves from the trees; and, indeed, this is so very little a matter to do, that it would be monstrous to suppose, that any farmer would neglect it on account of the labour or trouble; especially when we consider, *that the seed of two or three turnips is more than sufficient to sow an acre of land.* I, on one occasion, planted *twenty* turnips for seed, and the produce, besides what the little birds took as their share for having kept down the

caterpillars, was *twenty-two and a half pounds of clean seed.*

34. The sun is so ardent and the weather so fair here, compared with the drippy and chilly climate of England, while the birds here never touch this sort of seed, that a small plot of ground would, if well managed, produce a great quantity of seed. Whether it would *degenerate* is a matter that I have not *yet* ascertained; but which I am about to ascertain this year.

35. That all these precautions of *selecting the plants* and *transplanting* them are necessary, I know by experience. I, on one occasion, had sown all my own seed, and the plants had been carried off by the *fly*, of which I shall have to speak presently. I sent to a person who had raised some seed, which I afterwards found to have come from turnips, left promiscuous to go to seed in a part of a field where they had been sown. The consequence was, that a good *third part* of my crop had *no bulbs;* but consisted of a sort of *rape,* all leaves, and stalks growing very high. While even the rest of the crop bore no resemblance, either in point of size or of quality, to turnips, in the same field, from seed saved in a proper manner, though this latter was sown at a later period.

36. As to the *preserving* of the seed, it is an invariable rule applicable to all seeds, that seed,

kept in the pod to the very time of sowing, will vegetate more quickly and more vigorously than seed which has been some time threshed out. But, turnip seed will do very well, if threshed out as soon as ripe, and kept in a *dry place*, and not too much exposed to the air. A bag, hung up in a dry room, is the depository that I use. But, before being threshed out, the seed should be quite ripe, and, if cut off, or pulled up, which latter is the best way, before the pods are quite dead, the whole should be suffered to lie in the sun till the pods are perfectly dead, in order that the seed may imbibe its full nourishment, and come to complete perfection; otherwise the seed will *wither*, much of it will not grow at all, and that which does grow will produce plants inferior to those proceeding from well-ripened seed.

Time of Sowing.

37. Our *time of sowing* in England is from the first to the twentieth of June, though some persons sow in May, which is still better. This was one of the matters of the most deep interest with me, when I came to Hyde Park. I could not begin before the month of June; for I had no ground ready. But, then, I began with great care, on the second of June, sowing, in small plots, *once every week*, till the 30th of July. In *every case* the seed took well and the

plants grew well; but, having looked at the growth of the plots, first sown, and calculated upon the probable advancement of them, I fixed upon the 26*th of June* for the sowing of my principal crop.

38. I was particularly anxious to know, whether this country were cursed with the *Turnip Fly*, which is so destructive in England, It is a little insect about the size of a *bed flea*, and jumps away from all approaches exactly like that insect. It abounds, sometimes, in quantities so great as to eat up all the young plants, on hundreds and thousands of acres in a single day. It makes its attack when the plants are in the *seed-leaf;* and, it is so very generally prevalent, that it is always an even chance, at least, that every field that is sown will be thus wholly destroyed. There is no remedy but that of ploughing and sowing again; and this is frequently repeated *three times*, and even then, there is no crop. Volumes upon volumes have been written on the means of preventing, or mitigating, this calamity; but nothing effectual has ever been discovered; and, at last, the *only* means of *insuring* a crop of Ruta Baga in England, is, to raise the plants in small plots, sown at many different times, in the same manner as cabbages are sown, and, like cabbages, *transplant them;* of which mode of culture I shall speak by and by. It is very

singular, that a field, sown *one day*, wholly escapes, while a field, sown the *next day*, is wholly destroyed. Nay, a part of the same field, sown in the morning, will sometimes escape, while the part, sown in the afternoon, will be destroyed; and, sometimes the afternoon sowing is the part that is spared. To find a remedy for this evil has posed all the heads of all the naturalists and chemists of England. As an evil, the smut in wheat; the wire-worm; the grubs above-ground and under-ground; the caterpillars, green and black; the slug, red, black, and grey: though each a great tormentor, are nothing. Against all these there is *some remedy*, though expensive and plaguing; or, at any rate, their ravages are comparatively slow, and their *causes are known*. But, the ***Turnip Fly*** is the English farmer's evil genius. To discover a remedy for, or the cause of, this plague, has been object of enquiries, experiments, analyses, innumerable. Premium upon premium offered, has only produced pretended remedies, which have led to disappointment and mortification; and, I have no hesitation to say, that, if any man could find out a real remedy, and could communicate the means of cure, while he kept the nature of the means a secret, he would be much richer than he who should discover the longitude; for about *fifty thousand* farmers would very cheerfully pay him *ten guineas a year* each.

39. The reader will easily judge, then, of my anxiety to know, whether this mortal enemy of the farmer existed in Long Island. This was the first question which I put to every one of my neighbours, and I augured good, from their not appearing to understand what I meant. However, as my little plots of turnips came up successively, I watched them as our farmers do their fields in England. To my infinite satisfaction, I found that my alarms had been groundless. This circumstance, besides others that I have to mention by and by, gives to the stock-farmer in America so great an advantage over the farmer in England, or in any part of the middle and northern parts of Europe, that it is truly wonderful that the culture of this root has not, long ago, become general in this country.

40. The *time of sowing*, then, may be, as circumstances may require, *from the 25th of June to about the 10th of July*, as the result of my experiments will now show. The plants sown during the first fifteen days of June grew well, and attained great size and weight; but, though they did not actually *go off to seed*, they were very little short of so doing. They rose into large and long necks, and sent out sprouts from

the upper part of the bulb; and, then, the bulb itself (which is the thing sought after) swelled no more. The substance of the bulb became hard and stringy; and the turnips, upon the whole, were smaller and of greatly inferior quality, compared with those, which were sown at the proper time.

41. The turnips sown between the 15th and 26th of June had all these appearances and quality, only in a less degree. But, those which were sown on the 26th of June, were perfect in shape, size, and quality; and, though I have grown them larger in England, it was not done without more manure upon half an acre than I scratched together to put upon seven acres at Hyde Park; but of this I shall speak more particularly when I come to the *quantity of crop*.

42. The sowings which were made after the 26th of June, and before the 10th of July, did very well; and, one particular sowing on the 9th of July, on 12 rods, or perches, of ground, sixteen and a half feet to the rod, yielded 62 bushels, leaves and roots cut off, which is after the rate of 992 bushels to an acre. But this sowing was on ground extremely well prepared and sufficiently manured with ashes from *burnt earth;* a mode of raising manure of which I shall fully treat in a future chapter.

43. Though this crop was so large, sown on

the 9th of July, I would by no means recommend any farmer, who can sow sooner, to defer the business to that time; for, I am of opinion with the old folk in the West of England, that God is almost always on the side of *early* farmers. Besides, one delay too often produces another delay; and he who puts off to the 9th, may put off to the 19th.

44. The crops, in small plots, which I sowed after the 9th of July to the 30th of that month, *grew* very well; but they regularly succeeded each other in diminution of size; and, which is a great matter, the cold weather overtook them before they were *ripe;* and ripeness is full as necessary in the case of roots as in the case of apples or of peaches.

Quality and Preparation of the Seed.

45. As a fine, rich, loose garden mould, of great depth, and having a porous stratum under it, is best for every thing that vegetates, except plants that live best in water, so it is best for the Ruta Baga. But, I know of no soil in the United States, in which this root may not be cultivated with the greatest facility. A *pure sand*, or a *very stiff clay*, would not do well certainly; but I have never seen any of either in America. The soil that I cultivate is *poor*, almost proverbially; but, what it really is, is this: it is a light loam, approaching towards

the *sandy*. It is of a brownish colour about eight inches deep; then becomes more of a *red* for about another eight inches; and then comes a mixture of yellowish sand and of pebbles, which continues down to the depth of many feet.

46. So much for the *nature* of the land. As to its *state*, it was that of as complete *poverty* as can well be imagined. My main crop of Ruta Baga was sown upon two different pieces. One, of about three acres, had borne, in 1816, some Indian corn *stalks*, together with immense quantities of brambles, grass, and weeds, of all descriptions. The other, of about four acres, had, when I took to it, *rye* growing on it; but, this rye was so poor, that my neighbour assured me, that it could produce nothing, and he advised me to let the cattle and sheep take it for their trouble of walking over the ground, which advice I readily followed; but, when he heard me say, that I intended to sow Russia turnips on the same ground, he very kindly told me his opinion of the matter, which was, that I should certainly throw my labour wholly away.

47. With these two pieces of ground I went to work early in June. I ploughed them *very shallow*, thinking to drag the grassy clods up with the harrow, to put them in heaps and burn them, in which case I would (barring the *fly!*), have pledged my life for a crop of Ruta Baga.

It adversely happened to *rain*, when my clods should have been burnt, and the furrows were so solidly fixed down by the rain, that I could not tear them up with the harrow; and, besides, my *time of sowing* came on apace. Thus situated, and having no faith in what I was told about the *dangers of deep ploughing*, I fixed four oxen to a strong plough, and turned up soil that had not seen the sun for many, many long years. Another soaking rain came very soon after, and went, at once, to the bottom of my ploughing, instead of being carried away instantly by evaporation. I then harrowed the ground down level, in order to keep it *moist* as long as I could; for the sun now began to be the thing most dreaded.

48. In the meanwhile I was preparing my *manure*. There was nothing of the kind visible upon the place. But, I had the good luck to follow a person, who appears not to have known much of the use of *brooms*. By means of sweeping and raking and scratching in and round the house, the barn, the stables, the hen-roost, and the court and yard, I got together about *four hundred bushels* of not very bad turnip manure. This was not quite 60 bushels to an acre for my seven acres; or, *three gallons* to every square rod.

49. However, though I made use of these beggarly means, I would not be understood to

recommend the use of such means to others. On the contrary, I should have preferred good and clean land, and plenty of manure; but of this I shall speak again, when I have given an account of the manner of *sowing* and *transplanting*.

Manner of Sowing.

50. Thus fitted out with land and manure, I set to the work of sowing, which was performed, with the help of two ploughs and two pair of oxen, on the 25th, 26th, and 27th of June. The ploughmen put the ground up into little *ridges*, having *two furrows on each side of the ridge:* so that every ridge consisted of four furrows, or turnings over of the plough; and the tops of the ridges were about *four feet* from each other; and, as the ploughing was performed to a great depth, there was, of course, a very deep gutter between every two ridges.

51. I took care to have the manure placed so as to be *under the middle* of each ridge; that is to say, just beneath where my seed was to come. I had but a very small quantity of seed as well as of manure. This seed I had, however, brought from home, where it was raised by a neighbour, on whom I could rely, and I had no faith in any other. So that I was compelled to bestow it on the ridges with a very parsimonious hand; not having, I be-

lieve more than four pounds to sow on the seven acres. It was sown principally in this manner; a man went along by the side of each ridge, and put down two or three seeds in places at about *ten inches* from each other, just drawing a little earth over, and *pressing it on the seed*, in order to make it vegetate quickly before the earth became *too dry*. This is always a good thing to be done, and especially in dry weather, and under a hot sun. Seeds are very small things; and though, when we see them covered over with earth, we conclude that the earth must *touch them closely*, we should remember, that a very small cavity is sufficient to keep untouched nearly all round, in which case, under a hot sun, and near the surface, they are sure to perish, or, at least, to lie long, and until rain come, before they start.

52. I remember a remarkable instance of this in saving some turnips to transplant at Botley. The whole of a piece of ground was sown *broad-cast*. My gardener had been told to sow *in beds*, that we might go in to weed the plants; and, having forgotten this till after sowing, he clapped down his line, and divided the plot into beds by *treading very hard* a little path at the distance of every four feet. The weather was very dry and the wind very keen. It continued so for three weeks; and, at the end of that time, we had scarcely a turnip in the beds, where the

ground had been left raked over; but, in the *paths* we had an abundance, which grew to be very fine, and which, when transplanted, made part of a field which bore *thirty three tons to the acre*, and which, as a *whole field*, was the finest I ever saw in my life.

53. I cannot help endeavouring to press this fact upon the reader. Squeezing down the earth makes it touch the seed in all its parts, and then it will soon vegetate. It is for this reason, that barley and oat fields should be *rolled*, if the weather be dry; and, indeed, that all seeds should be pressed down, if the state of the earth will admit of it.

54. This mode of sowing is neither tedious nor expensive. Two men sowed the whole of my seven acres in the three days, which, when we consider the value of the crop, and the saving in the after-culture, is really not worth mentioning. I do not think, that any sowing by drill is so good, or, in the end, so cheap as this. Drills miss very often in the sowings of such small seeds. However, the thing may be done by hand in a less precise manner. One man would have sown the seven acres in a day, by just scattering the seeds along on the top of the ridge, where they might have been buried with the rake, and pressed down by a spade or shovel or some other flat instrument. A slight roller to take two ridges at once, the horse

walking in the gutter between, is what I used to make use of when I sowed on ridges; and, who can want such a roller in America, as long as he has an axe and an auger in his house? Indeed, this whole matter is such a trifle, when compared with the importance of the object, that it is not to be believed, that any man will think it worth the smallest notice as counted amongst the means of obtaining that object.

55. *Broad cast sowing* will, however, probably, be, in most cases, preferred; and, this mode of sowing is pretty well understood from general experience. What is required here, is, that the ground be well ploughed, finely harrowed, and the seeds thinly and evenly sown over it, to the amount of about two pounds of seed to an acre; but, then, if the weather be dry, the seed should, by all means, be *rolled* down. When I have spoken of the *after-culture*, I shall compare the two methods of sowing, the *ridge* and the *broad-cast*, in order that the reader may be the better able to say, which of the two is entitled to the preference.

After-culture.

56. In relating what I did in this respect, I shall take it for granted, that the reader will understand me as describing what I think ought to be done.

57. When my ridges were laid up, and my

seed was sown, my neighbours thought, that there was an end of the process; for, they all said, that, if the seed ever came up, being upon those high ridges, the plants never could live under the scorching of the sun. I knew that this was an erroneous notion; but I had not much confidence in the powers of the soil, it being so evidently poor, and my supply of manure so scanty.

58. The plants, however, made their appearance with great regularity; no *fly* came to annoy them. The moment they were fairly up, we went with a very small hoe, and took all but one in each ten or eleven or twelve inches, and thus left them singly placed. This is a great point; for they begin to rob one another at a very early age; and, if left two or three weeks to rob each other, before they are set out singly, the crop will be diminished one-half. To set the plants out in this way was a very easy and quickly-performed business; but, it is a business to be left to no one but a careful man. Boys can never safely be trusted with the deciding, at discretion, whether you shall have a large crop or a small one.

59. But, now, something else began to appear as well as turnip-plants; for, all the long grass and weeds having dropped their seeds the summer before, and, probably, for many summers, they now came forth to demand their share of

that nourishment, produced by the fermentation, the dews, and particularly the *sun*, which shines on all alike. I never saw a fiftieth part so many weeds in my life upon a like space of ground. Their little seed leaves, of various hues, formed a perfect mat on the ground. And now it was, that my *wide ridges*, which had appeared to my neighbours to be so very singular and so unnecessary, were *absolutely* necessary. First we went with a hoe, and hoed the *tops of the ridges*, about six inches wide. There were all the plants, then, clear and clean at once, with an expense of about half a day's work to an acre. Then we came, in our Botley fashion, with a single horse-plough, took a furrow from the side of one ridge going up the field, a furrow from the other ridge coming down, then another furrow from the same side of the first ridge going up, and another from the same side of the other ridge coming down. In the taking away of the last two furrows, we went within *three inches* of the turnip-plants. Thus there was a ridge over the original gutter. Then we turned these furrows *back again* to the turnips. And, having gone, in this manner, over the whole piece, there it was with not a weed alive in it. All killed by the sun, and the field as clean and as fine as any garden that ever was seen.

60. Those who know the effect of *tillage be-*

tween growing plants, and especially if the earth be *moved deep* (and, indeed, what American does not know what such effect is, seeing that, without it, there would be no Indian Corn?) those that reflect on this effect, may guess at the effect on my Ruta Baga plants, which soon gave me, by their appearance, a decided proof, that TULL's principles are always true, in whatever soil or climate applied.

61. It was now a very beautiful thing to see a regular, unbroken line of fine, fresh-looking plants upon the tops of those wide ridges, which had been thought to be so very whimsical and unnecessary. But, why have the ridges *so very wide?* This question was not new to me, who had to answer it a thousand times in England. It is because you cannot plough *deep* and *clean* in a narrower space than four feet; and, it is the deep and clean ploughing that I regard as the surest means of a large crop, especially in poor, or indifferent ground. It is a great error to suppose, that there is any ground *lost* by these wide intervals. My crop of *thirty-three tons*, or *thirteen hundred and twenty bushels*, to the acre, taking a whole field together, had the same sort of intervals; while my neighbour's with two feet intervals, never arrived at two-thirds of the weight of that crop. There is no *ground lost;* for, any one, who has a mind to do it, may satisfy himself, that the *lateral roots*

of any fine large turnip will extend more than *six feet* from the bulb of the plant. The intervals are full of these roots, the breaking of which and the moving of which, as in the case of Indian Corn, gives new food and new roots, and produces wonderful effects on the plants. Wide as my intervals were, the leaves of some of the plants very nearly touched those of the plants on the adjoining ridge, before the end of their growth; and I have had them frequently meet in this way in England. They would always do it here, if the ground were rich and the tillage proper. How, then, can the intervals be too wide, if the plants occupy the interval? And how can any ground be lost if every inch be full of roots and shaded by leaves?

62. After the last-mentioned operation my plants remained till the weeds had again made their appearance; or, rather, till a new brood had started up. When this was the case, we went with the hoe again and cleaned the tops of the ridges as before. The weeds under this all-powerful sun, instantly perish. Then we repeated the former operation with the one-horse plough. After this nothing was done but to pull up now and then a weed, which had escaped the hoe; for, as to the plough-share, nothing escapes that.

63. Now, I think, no farmer can discover in

this process any thing more difficult, more troublesome, more expensive, than in the process absolutely necessary to the obtaining of a crop of Indian Corn. And yet, I will venture to say, that in any land, capable of bearing *fifty bushels* of corn upon an acre, more than a thousand bushels of Ruta Baga may, in the above described manner, be raised.

64. In the *broad-cast* method the after-culture must, of course, be confined to *hoeing*, or, as TULL calls it, *scratching*. In England, the hoer goes in when the plants are about four inches high, and hoes all the ground, setting out the plants to about *eighteen inches* apart; and, if the ground be at all foul, he is obliged to go in again in about a month afterwards; to hoe the ground again. This is all that is done; and a very poor all it is, as the crops, on the very best ground, compared with the ridged crops, invariably show.

Transplanting.

65. This is a third mode of cultivating the RUTA BAGA; and, in certain cases, far preferable to either of the other two. My *large* crops at Botley were from roots *transplanted.* I resorted to this mode in order to insure a crop in spite of the *fly;* but, I am of opinion, that it is, in all cases, the best mode, provided *hands* can be obtained in sufficient number, just for a few

days, or weeks, as the quantity may be, when the land and the plants are ready.

66. Much light is thrown on matters of this sort by describing what one has *done one's self* relating to them. This is practice at once; or, at least, it comes much nearer to it than any instructions possibly can.

67. It was an accident that led me to the practice. In the summer of 1812, I had a piece of *Ruta Baga* in the middle of a field, or, rather, the piece occupied a part of the field, having a crop of carrots on one side and a crop of Mangel Wurzel on the other side. On the 20th of July the turnips, or rather, those of them which had escaped the fly, began to grow pretty well. They had been sown in drills; and I was anxious to fill up the spaces, which had been occasioned by the ravages of the *fly*. I, therefore, took the supernumerary plants, which I found in the un-attacked places, and filled up the rows by transplantation, which I did also in two other fields.

68. The turnips, thus transplanted, *grew*, and, in fact, were pretty good; but, they were very far inferior to those which had retained their original places. But, it happened, that on one side of the above-mentioned piece of turnips, there was a vacant space of about a yard in breadth. When the ploughman had finished ploughing between the rows of turnips, I made

him plough up that spare ground very deep, and upon it I made my gardener go and plant two rows of turnips. These became the largest and finest of the whole piece, though transplanted two days later than those which had been transplanted in the rows throughout the piece. The cause of this remarkable difference I, at once, saw, was, that these had been put into *newly-ploughed* ground; for, though I had not read much of TULL at the time here referred to, I knew, from the experience of my whole life, that plants as well as seeds ought always to go into ground as recently moved as possible; because at every moving of the earth, and particularly at every turning of it, a new process of fermentation takes place, fresh exhalations arise, and a supply of the *food of plants* is thus prepared for the newly arrived guests. Mr. CURWEN, the Member of Parliament, though a poor thing as to public matters, has published not a bad book on *agriculture*. It is not bad, because it contains many authentic accounts of experiments made by himself; though I never can think of his book without thinking, at the same time, of the gross and scandalous plagiarisms, which he has committed upon TULL. Without mentioning particulars, the "*Honourable* Member" will, I am sure, know what I mean, if this page should ever have the honour to fall under his eye; and he will, I hope, repent, and give proof of his repentance, by a restoration of the property to the right owner.

69. However, Mr. CURWEN, in his book, gives an account of the wonderful *effects of moving the ground* between plants in rows; and he tells us of an experiment, which he made, and which proved, that from ground just ploughed, in a very dry time, an *exhalation* of *many tons* weight, per acre, took place, during the first twenty-four hours after ploughing, and of a less and less number of tons, during the three or four succeeding twenty-four hours; that, in the course of about a week, the exhalation *ceased;* and that, during the whole period, the ground, though in the *same field,* which had *not been ploughed* when the other ground was, exhaled *not an ounce!* When I read this in Mr. CURWEN's book, which was *before* I had read TULL, I called to mind, that, having once dug the ground between some rows of *part* of a plot of cabbages in my garden, in order to plant some late peas, I perceived (it was in a dry time) the cabbages, the next morning, in the part recently dug, with big *drops of dew* hanging on the edges of the leaves, and in the other, or undug part of the plot, no drops at all. I had forgotten the fact till I read Mr. CURWEN, and I never knew the *cause* till I read the real *Father of English Husbandry.*

70. From this digression I return to the his-

tory, first of my English transplanting. I saw, at once, that the only way to ensure a crop of turnips was by *transplantation.* The next year, therefore, I prepared a field of *five acres,* and another of *twelve.* I made *ridges,* in the manner described, for sowing; and, on the 7th of June in the first field, and on the 20th of July in the second field, I planted my plants. I ascertained to an exactness, that there were *thirty-three tons to an acre,* throughout the whole seventeen acres. After this, I never used any other method. I never *saw* above *half* as great a crop in any other person's land; and, though we read of much greater in *agricultural prize reports,* they must have been of the extent of a *single acre,* or something in that way. In my usual order, the ridges four feet asunder, and the plants *a foot* asunder on the ridge, there were *ten thousand eight hundred and thirty turnips* on the acre of ground; and, therefore, for an acre to weigh *thirty-three tons,* each turnip must weigh very nearly *seven pounds.* After the time here spoken of, I had an acre or two at the end of a large field, transplanted on the 13th of July, which, probably, weighed *fifty tons an acre.* I delayed to have them weighed till a fire happened in some of my farm buildings, which produced a further delay, and so the thing was not done at all; but, I weighed *one waggon load,* the turnips of which *averaged eleven pounds each;* and several weighed fourteen pounds each. My very largest upon Long Island weighed *twelve pounds and a half.* In all these cases, as well here as in England, the produce was from *transplanted* plants; though at Hyde Park, I have many turnips of more than ten pounds weight each from *sown* plants, some of which, on account of the great perfection in their qualities, I have selected, and am now planting out, for seed.

71. I will now give a full account of my transplanting at Hyde Park. In a part of the ground, which was put into *ridges* and sown, I scattered the seed along very thinly upon the top of the ridge. But, however thinly you may attempt to *scatter* such small seeds, there will always be too many plants, if the tillage be good and the seed good also. I suffered these plants to stand as they came up; and, they stood much too long, on account of my want of hands, or, rather, my want of time to attend to give my directions in the transplanting; and, indeed, my *example* too; for, I met not with a man who knew how to *fix* a plant in the ground; and, strange as it may appear, more than half the bulk of crop depends on a little, trifling, contemptible twist of the *setting-stick,* or *dibble;* a thing very well known to all gardeners in the case of cabbages, and about which, therefore, I will give, by and by, very plain instructions.

72. Thus puzzled, and not being able to spare time to do the job myself, I was one day looking at my poor plants, which were daily suffering for want of removal, and was thinking how glad I should be of one of the CHURCHERS at Botley, who, I thought to myself, would soon clap me out my turnip patch. At this very time, and into the field itself, came a cousin of one of these CHURCHERS, who had lately arrived from England! It was very strange, but literally the fact.

73. To work Churcher and I went, and, with the aid of persons to pull up the plants and bring them to us, we planted out about two acres, in the *mornings and evenings* of six days; for the weather was too hot for us to keep out *after breakfast*, until about two hours before sun-set. There was a friend staying with me, who helped us plant, and who did, indeed, as much of the work as either Churcher or I.

74. The *time* when this was done was from the 21st to the 28th of *August*, one Sunday and one day of no planting, having intervened. Every body knows, that this is the very *hottest* season of the year; and, as it happened, this was, last summer, *the very driest* also. The weather had been hot and dry from the *10th of August*; and so it continued to the *12th of September*. Any gentleman who has kept a journal of last year, upon Long Island, will

know this to be correct. Who would have thought to see these plants thrive? who would have thought to see them *live?* The next day after being planted, their leaves crumbled between our fingers like the old leaves of trees. In two days there was no more appearance of a crop upon the ground than there was of a crop on the Turnpike-road. But, on the 2nd of September, as I have it in my memorandum book, the plants *began to show life;* and, before the rain came, on the 12th, the piece began to have an air of verdure, and, indeed, to grow and to promise a good crop.

75. I will speak of the *bulk* of this crop by and by; but, I must here mention another transplantation that I made in the latter end of *July*. A plot of ground, occupied by one of my earliest sowings, had the turnips standing in it in rows at eighteen inches asunder, and at a foot asunder in the rows. Towards the middle of July I found, that one half of the rows must be taken away, or that the whole would be of little value. Having pulled up the plants, I intended to translate them (as they say of Bishops) from the garden to the field; but, I had no ground ready. However, I did not like to throw away these plants, which had already bulbs as large as hens' eggs. They were carried into the cellar, where they lay in a heap, till (which would soon happen in such hot weather)

they began to *ferment*. This made the most of their leaves turn white. Unwilling, still, to throw them away, I next laid them *on the grass* in the front of the house, where they got the dews in the night, and they were covered with a mat during the day, except two days, when they were overlooked, or, rather, neglected. The heat was very great, and, at last, supposing these plants *dead*, I did not cover them any more. There they lay abandoned till the 24th of July, on which day I began planting *Cabbages* in my field. I then thought, that I would *try* the hardiness of a *Ruta Baga plant*. I took these same abandoned plants, without a morsel of *green* left about them; planted them in part of a row of the piece of cabbages; and they, *a hundred and six* in number, weighed, when they were taken up, in December, *nine hundred and one pounds*. One of these turnips weighed *twelve* pounds and a half.

76. But, it ought to be observed, that this was in ground which had been got up in my best manner; that it had some of the best of my manure; and, that uncommon pains were taken by myself in the putting in of the plants. This experiment shows, what a hardy plant this is; but, I must caution the reader against a belief, that it is either desirable or prudent to put this quality to so severe a test. There is no necessity for it, in general; and, indeed, the

rule is, that the shorter time the plants are out of the ground the better.

77. But, as to the business of transplanting, there is one very material observation to make. The ground ought to be as *fresh*; that is to say, as *recently moved* by the plough, as possible; and that for the reasons before stated. The way I go on is this: my land is put up into ridges, as described under the head of *manner of sowing*. This is done before-hand, several days; or, it may be, a week or more. When we have our plants and hands all ready, the ploughman begins, and *turns* in the ridges; that is to say, ploughs the ground back again, so that the top of the new ploughed ridge stands over the place where the channel, or gutter, or deep furrow, was, before he began. As soon as he has finished the first ridge, the planters plant it, while he is ploughing the second: and so on throughout the field. That this is not a very tedious process the reader needs only to be told, that, in 1816, I had *fifty-two acres* of Ruta Baga planted in this way; and I think I had more than *fifty thousand bushels*. A smart hand will plant half an acre a day, with a girl or a boy to drop the plants for him. I had a man, who planted an acre a day many a time. But, supposing that a quarter of an acre is a day's work, what are *four days' work*, when put in competition with the value of an acre of this invaluable root?

And what farmer is there, who has common industry, who would grudge to bend his *own* back eight or twelve days, for the sake of keeping all his stock through the Spring months, when dry food is loathsome to them, and when grass is by nature denied?

78. Observing well what has been said about earth *perfectly fresh*, and never forgetting this, let us now talk about the *act* of planting; the mere mechanical operation of putting the plant into the ground. We have a *setting-stick*, which should be the top of a spade-handle cut off, about ten inches below the eye. It must be pointed smoothly; and, if it be shod with thin iron; that is to say, covered with an iron sheath, it will work more smoothly, and do its business the better. At any rate the point should be nicely smoothed, and so should the whole of the tool. The planting is performed like that of cabbage-plants; but, as I have met with very few persons, out of the market gardens, and gentlemen's gardens in England, who knew how to plant a cabbage-plant, so I am led to suppose, that very few, comparatively speaking, know how to plant a turnip-plant.

79. You constantly hear people say, that they *wait for a shower*, in order to put out their cabbage-plants. Never was there an error more general or more complete in all its parts. Instead of rainy weather being the best time, it is

the very worst time, for this business of transplantation, whether of cabbages or of any thing else, from a lettuce-plant to an apple-tree. I have proved the fact, in scores upon scores of instances. The first time that I had any experience of the matter was in the planting out of a plot of cabbages in my garden at Wilmington in Delaware. I planted in dry weather, and, as I had always done, in such cases, I *watered* the plants heavily; but, being called away for some purpose, I left one row *unwatered*, and it happened, that it so continued without my observing it till the next day. The sun had so completely scorched it by the next night, that, when I repeated my watering of the rest, I left it, as being unworthy of my care, intending to plant some other thing in the ground occupied by this *dead* row. But, in a few days, I saw, that it was not dead. It grew soon afterwards; and, in the end, the cabbages of my *dead* row were not only larger, but *earlier* in loaving, than any of the rest of the plot.

80. The reason is this: if plants are put into *wet* earth, the setting-stick squeezes the earth up against the tender fibres in a *mortar-like* state. The sun comes and bakes this mortar into a sort of glazed clod. The hole made by the stick is also a *smooth* sided hole, which *retains its form*, and presents, on every side, an impenetrable substance to the fibres. In short,

such as the hole is made, such it, in a great measure, remains, and the roots are cooped up in this sort of *well*, instead of having a free course left them to seek their food on every side. Besides this, the fibres get, from being wet when planted, into a small compass. They all cling about the tap root, and are stuck on to it by the wet dirt; in which state, if a hot sun follow, they are all baked together in a lump, and cannot stir. On the contrary, when put into ground *unwet*, the reverse of all this takes place; and the *fresh* earth will, under *any sun*, supply moisture in quantity sufficient.

81. Yet, in July and August, both in England and America, how many thousands and thousands are *waiting for a shower* to put out their plants! And then, when the long-wished-for shower comes, they must plant upon *stale* ground, for they have it dug ready, as it were, for the purpose of keeping them company in waiting for the shower. Thus all the fermentations, which took place upon the digging, is gone; and, when the planting has once taken place, farewell to the spade! For, it appears to be a *privilege* of the Indian corn to receive something like good usage *after being planted.* It is very strange that it should have been thus, for what *reason* is there for other plants not enjoying a similar benefit? The reason is, that they will produce *something* without it; and

the Indian corn will positively produce *nothing;* for which the Indian corn is very much to be commended. As an instance of this effect of deeply moving the earth between growing crops, I will mention, that, in the month of June, and on the 26th of that month, a very kind neighbour of mine, in whose garden I was, showed me a plot of *Green Savoy Cabbages,* which he had planted in some ground as rich as ground could be. He had planted them about three weeks before; and they appeared very fine indeed. In the seed bed, from which he had taken his plants, there remained about *a hundred;* but, as they had been left as of *no use,* they had drawn each other up, in company with the weeds, till they were about eighteen inches high, having only a starved leaf or two upon the top of each. I asked my neighbour to give me these plants, which he readily did; but begged me not to plant them, for, he assured me, that they would *come to nothing.* Indeed, they were a ragged lot; but, I had no plants of my own sowing more than two inches high. I, therefore, took these plants and dug some ground for them between some rows of scarlet blossomed beans, which mount upon poles. I cut a stick on purpose, and put the plants very deep into the ground. My beans came off in August, and then the ground was well dug between the rows of cabbages. In September, mine had far surpassed the prime plants of my neighbour. And, in the end I believe, that ten of *my cabbages* would have weighed *a hundred* of his, leaving out the stems in both cases. But, his had remained uncultivated *after planting*. The ground, battered down by successive rains, had become hard as a brick. All the stores of food had been locked up, and lay in a dormant state. There had been no renewed fermentations, and no exhalations.

82. Having now said what, I would fain hope, will convince every reader of the folly of *waiting for a shower* in order to transplant plants of any sort, I will now speak of the mere *act* of planting, more particularly than I have hitherto spoken.

83. The hole is made sufficiently deep; deeper than the length of the root does really require; but, the root should not be *bent* at the point, if it can be avoided. Then, while one hand holds the plant, with its root in the hole, the other hand applies the setting-stick to the earth on one side of the hole, the stick being held in such a way as to form a sharp triangle with the plant. Then pushing the stick down, so that *its point goes a little deeper than the point of the root*, and giving it a little *twist*, it presses the earth against the *point*, or *bottom*, of the root. And thus all is safe, and the plant is sure to grow.

84. The general, and almost universal fault, is, that the planter, when he has put the root into the hole, draws the earth up against the *upper part* of the root, or stem, and, if he presses pretty well there, he thinks that the planting is well done. But, it is the *point* of the root, against which the earth ought to be pressed, for there the *fibres* are; and, if they do not *touch* the earth *closely*, the plant will not thrive. The *reasons* have been given in paragraphs 51 and 52, in speaking of the sowing of seeds. It is the same in all cases of *transplanting* or *planting*. Trees, for instance, will be sure to grow, if you *sift* the earth, or pulverize it very finely, and place it carefully and closely about the roots. When we plant a tree, we see all *covered* by tumbling in the earth; and, it appears whimsical to suppose, that the earth does *not touch* all the roots. But, the fact is, that unless great pains be taken, there will be many cavities in the hole where the tree is planted; and, in whatever places the earth does not closely touch the root, the root will *mould*, become cankered, and will lead to the producing of a poor tree.

85. When I began transplanting in fields in England, I had infinite difficulty in making my planters attend to the directions, which I have here given. "*The point of the stick to the point of the root!*" was my constant cry. As

I could not be much with my work-people, I used, in order to try whether they had planted properly, to go after them, and now-and-then take the tip of a leaf between my finger and thumb. If the plant resisted the pull, so as for the bit of leaf to come away, I was sure that the plant was well fixed; but, if the pull brought up the plant out of the ground; then I was sure, that the planting was not well done. After the first field or two, I had no trouble. My work was as well done, as if the whole had been done by myself. My planting was done chiefly by *young women*, each of whom would plant half an acre a day, and their pay was *ten pence sterling* a day. What a shame, then, for any *man* to shrink at the *trouble* and *labour* of such a matter! Nor, let it be imagined, that these young women were poor, miserable, ragged, squalid creatures. They were just the contrary. On a Sunday they appeared in their *white* dresses, and with silk umbrellas over their heads. Their constant labour afforded the means of dressing well, their early rising and exercise gave them health, their habitual cleanliness and neatness, for which the women of the South of England are so justly famed, served to aid in the completing of their appearance, which was that of fine rosy-cheeked country-girls, fit to be the helpmates, and not a burden, of their future husbands.

86. But, at any rate, what can be said for a *man* that thinks too much of such a piece of labour? The earth is extremely grateful; but it must and will have something to be grateful for. As far as my little experience has enabled me to speak, I find no want of *willingness to learn* in any of the American workmen. Ours, in England, are apt to be very *obstinate*, especially if getting a little old. They do not like to be *taught* any thing. They say, and they think, that what their fathers did was best. To tell them, that it was *your* affair, and not *theirs*, is nothing. To tell them, that the loss, if any, will fall upon *you*, and not upon *them*, has very little weight. They argue, that, they being the real *doers*, ought to be the best judges of the *mode of doing*. And, indeed, in *most cases*, they are, and go about this work with wonderful skill and judgment. But, then, it is so difficult to induce them cordially to do any thing *new*, or any old thing in a *new way;* and the abler they are as workmen, the more untractable they are, and the more difficult to be persuaded, that any one knows any thing, relating to farming affairs, better than they do. It was this difficulty that made me resort to the employment of young women in the most important part of my farming, the providing of immense quantities of cattle-food. But, I do not find this difficulty here, where no workmen are ob-

stinate, and where, too, all one's neighbours *rejoice at one's success*, which is by no means the case amongst the farmers in England.

87. Having now given instructions relative to the business of *transplanting* of the *Ruta Baga*, let us see, whether it be not preferable to either the *ridge-sowing* method, or the *broad-cast* method.

88. In the first place, when the seed is sown on the ground where the plants are to come to perfection, the ground, as we have seen in paragraph 40 and paragraph 47, must be prepared *early in June*, at the latest; but, in the transplanting method, this work may be put off, if need be, till *early in August*, as we have seen in paragraphs 74 and 75. However, the best time for transplanting is about the 26th of July, and this gives a *month* for preparation of land, more than is allowed in the sowing methods. This, of itself, is a great matter; but, there are others of far greater importance.

89. This transplanted crop may follow *another crop on the same land.* Early cabbages will loave and be away; early peas will be ripe and off; nay, even wheat, and all grain, except buckwheat, may be succeeded by Ruta Baga transplanted. I had crops to succeed *Potatoes, Kidney Beans, White Peas, Onions,* and even *Indian Corn,* gathered to eat green; and, the reader will please to bear in mind, that I did

not sow, or plant, any of my *first* crops, just mentioned, till the *month of June.* What might a man do, then, who is in a state to begin with his first crops as soon as he pleases! Who has his land all in order, and his manure ready to be applied.

90. Another great advantage of the transplanting method is, that it saves almost the whole of the *after-culture.* There is *no hoeing;* no *thinning* of the plants; and not more than one ploughing between the ridges. This is a great consideration, and should always be thought of, when we are talking of the *trouble* of transplanting. The turnips which I have mentioned in paragraphs 72 and 73 had *no after-culture* of any sort; for they soon spread the ground over with their leaves; and, indeed, after July, very few weeds make their appearance. The season for their coming up is passed; and, as every farmer well knows, if there be no weeds up at the end of July, very few will come that summer.

91. Another advantage of the transplanting method is, that you are *sure* that you have your right number of plants, and those regularly placed. For, in spite of all you can do in sowing, there will be deficiencies and irregularities. The seed may not come up, in some places. The plants may, in some places, be destroyed in their infant state. They may, now

and then, be cut off with the hoe. The best plants may sometimes be cut up, and the inferior plants left to grow. And, in the broadcast method, the irregularity and uncertainty must be obvious to every one. None of these injurious consequences can arise in the transplanting method. Here, when the work is once well done, the crop is certain, and all cares are at an end.

92. In taking my leave of this part of my treatise, I must observe, that it is useless, and, indeed, unjust, for any man to expect success, unless he *attend* to the thing *himself*, at least, till he has made the matter perfectly familiar to his work-people. To neglect *any part* of the business is, in fact, to neglect the whole; just as much as neglecting to put up one of the sides of a building, is to neglect the whole building. Were it a matter of trifling moment, personal attention might be dispensed with; but, as I shall, I think, clearly show, this is a matter of very great moment to every farmer. The object is, not merely to get roots, but to get them of a *large size;* for, as I shall show, there is an amazing difference in this. And, large roots are not to be gotten without *care*, which, by the by, *costs nothing*. Besides, the care bestowed in obtaining this crop, removes all the million of cares and vexations of the Spring months, when bleatings everlasting din

the farmer almost out of his senses, and make him ready to knock the brains out of the clamorous flock, when he ought to feel pleasure in the filling of their bellies.

93. Having now done with the different modes of *cropping* the ground with Ruta Baga, I will, as I proposed in paragraph 49, speak about the *preparation of the land generally;* and in doing this, I shall suppose the land to have borne a good crop of wheat the preceding year, and, of course, to be in *good heart,* as we call it in England.

94. I would plough this ground in the fall into ridges four feet asunder. The ploughing should be very deep, and the ridges well laid up. In this situation it would, by the successive frosts and thaws, be shaken and broken fine as powder by March or April. In April, it should be turned back; always ploughing *deep.* A crop of weeds would be well set upon it by the first of June, when they should be smothered by another turning back. Then, about the *third week* in June, I would carry in my manure, and fling it along on the trenches or furrows. After this I would follow the turning back for the sowing, as is directed in paragraph 50. Now, here are *four ploughings.* And what is the *cost* of these ploughings? My man, a black man, a native of this Island, ploughs with his pair of oxen and no driver *an acre and*

a half a day, and his oxen keep their flesh extremely well upon the *refuse* of the ***Ruta Baga*** which I send to market. What is the *cost* then? And, what a fine state the grass is thus brought into! A very different thing indeed is it to plough hard ground, from what it is to plough ground in this fine, broken state. Besides, every previous ploughing, especially *deep* ploughing, is equal to a seventh part of an ordinary coat of manure.

95. In the broad-cast method I would give the same number of previous ploughings, and at the same seasons of the year. I would spread the manure over the ground just before I ploughed it for sowing. Then, when I ploughed for the sowing, I would, if I had only one pair of oxen, plough about half an acre, harrow the ground, sow it *immediately*, and roll it with a *light* roller, which a little horse might draw, in order to *press the earth* about the seeds, and cover them too. There need be no *harrowing after sowing*. We never do it in England. The roller does all very completely, and the sowing upon the *fresh earth* will, under any sun, furnish the moisture sufficient. I once sowed, on ridges, with a BENNETT's drill, and neither harrowed nor rolled, nor used any means at all of covering the seed; and yet I had plenty of plants and a very fine crop of turnips. I sowed a piece of white turnips, broad-cast, at Hyde Park,

last summer, on the eleventh of August, which did very well, though neither harrowed nor rolled after being sown. But, in both these cases, there came rain directly after the sowing, which battered down the seeds; and which rain, indeed, it was, which prevented the rolling; for, that cannot take place when the ground is *wet*; because, then, the earth will adhere to the roller, which will go on growing in size like a rolling snow-ball. To harrow after the sowing is sure to do mischief. We always bury seeds *too deep*; and, in the operation of harrowing, more than half the seeds of turnips must be destroyed, or rendered useless. If a seed lies beyond the proper depth, it will either remain in a quiescent state, until some movement of the earth bring it up to the distance from the surface, which will make it vegetate, or, it will vegetate, and come up *later* than the rest of the plants. It will be *feebler* also; and it will never be equal to a plant, which has come from a seed near the surface.

96. Before I proceed further, it may not be amiss to say something more respecting the *burying* of seed, though it may here be rather out of place. Seeds buried below their proper depth, do not *come up*; but, many of them are near enough to the surface, sometimes, to *vegetate*, without coming up; and then they die. This is the case, in many instances, with more than

one half of the seed that is sown. But, if seeds be buried so deep, that they *do not even vegetate*, then they do not die; and this is one cause, though not the only cause, of our wondering to see *weeds* come up, where we are sure that no seeds have fallen for many years. At every digging, or every ploughing, more or less of the seeds, that have formerly been buried, come up near the surface; and then they vegetate. I have seen many instances in proof of this fact; but, the particular instance, on which I found the positiveness of my assertion, was in *Parsnip* seed. It is a very delicate seed. It will, if beat out, keep only *one year*. I had a row of fine seed parsnips in my garden, many of the seeds of which fell in the gathering. The ground was dug in the fall; and, when I saw it full of parsnips in the Spring, I only regarded this as a proof, that parsnips *might* be sown in the fall, though I have since proved, that is a very bad practice. The ground was dug again, and again for several successive years; and there was *always a crop of parsnips*, without a grain of seed ever having been sown on it. But lest any one should take it into his head, that this is a most delightful way of saving the trouble of sowing, I ought to state, that the parsnips coming thus at random, gave me a great deal more labour, than the same crop would have given me in the regular way of sowing. Besides, the fall is not the time to sow, as my *big* and white parsnips, now selling in New York market, may clearly show; seeing that *they* were sown in *June!* And yet, people are flocking to the *Western Countries* in search of *rich land*, while thousands of acres of such land as I occupy are lying waste in Long Island, within three hours drive of the all-consuming and incessantly increasing city of New York!

97. I have now spoken of the preparation of the land for the reception of *seeds*. As to the preparation in the case of *transplantation*, it might be just the same as for the sowing on ridges. But here might, in this case, be one more *previous* ploughing, always taking care to plough in *dry weather*, which is an observation I ought to have made before.

98. But, why should not the plants, in this case, *succeed some other good crop*, as mentioned before? I *sowed* some early peas (brought from England) on the 2nd of June. I *harvested* them, quite ripe and hard, on the 31st of July; and I had very fine Ruta Baga, some weighing six pounds each, after the peas. How little is known of the powers of this soil and climate! My potatoes were of the *kidney sort*, which, as every one knows, is not an *early* sort. They were planted on the 2nd of June; and they were succeeded by a most abundant crop of Ruta Baga. And, the manure for the peas and

potatoes served for the Ruta Baga also. In surveying my crops and feeling grateful to the kind earth and the glorious sun that produce these, to me, most delightful objects, how often have I turned, with an aching heart, towards the ill-treated Englishmen, shut up in dungeons by remorseless tyrants, while not a word had been uttered in their defence by, and while they were receiving not one cheering visit, or comforting word from, SIR FRANCIS BURDETT, who had been the great immediate cause of their incarceration!

99. As to the quantity and sort of *manure* to be used in general, it may be the same as for a sowing of rye, or of wheat. I should prefer *ashes*; but, my large crops in England were on yard-dung, first thrown into a heap, and afterwards *turned* once or twice, in the usual manner as practised in England. At Hyde Park I had nothing but *rakings up* about the yard, barn, &c. as described before. What I should do, and what I shall do this year, is, to *make ashes* out of *dirt*, or *earth*, of any sort, not very stony. Nothing is so easy as this, especially in this fine climate. I see people go with their waggons five miles for *soaper's ashes*; that is to say, *spent* ashes, which they purchase at the landing place (for they come to the island in vessels) at the rate of about five dollars for forty bushels. Add the expense of land-carriage, and the forty

bushels do not cost less than *ten dollars.* I am of opinion, that, by the burning of *earth,* as much manure may be got upon the land for *half a dollar.* I made an experiment last summer, which convinces me, that, if the spent ashes be received as *a gift* at *three miles* distance of land-carriage, they are not a gift worth accepting. But, this experiment was upon a small scale; and, therefore, I will not now speak positively on the subject.

100. I am now preparing to make a *perfect* trial of these ashes. I have just ploughed up a piece of ground, in which, a few years ago, Indian Corn was planted, and produced, as I am assured, only *stalks,* and those not more than *two feet high.* The ground has, every year since, borne a crop of weeds, rough grass, and briars, or brambles. The piece is about *ten acres.* I intend to have Indian corn on it; and, my manure shall be *made* on the spot, and consist of nothing but *burnt earth.* If I have a decent crop of Indian corn on this land so manured, it will, I think, puzzle my good neighbours to give a good reason for their *going five miles for spent ashes.*

101. Whether I succeed, or not, I will give an account of my experiment. This I know, that I, in the year 1815, burnt ashes, in one heap, to the amount of about *two hundred* English cart-loads, each load holding about

forty bushels. I should not suppose, that the burning cost me more than *five dollars;* and there they were upon the spot, in the very field, where they were used. As to their effect, I used them for the transplanted Ruta Baga and Mangel Wurzel, and they produced full as great an effect as the yard-dung used on the same land. This process of burning earth into ashes, *without suffering the smoke to escape,* during any part of the process, is a discovery of *Irish* origin. It was pointed out to me by Mr. WILLIAM GAUNTLETT of Winchester, late a Commissary with the army in Spain. To this gentleman I also owe, England owes, and I hope America will owe, the *best sort of hogs,* that are, I believe, in the world. I was wholly unacquainted with Mr. GAUNTLETT, till the summer of 1815, when, happening to pass by my farm, he saw my hogs, cows, &c. and, when he came to my house he called, and told me, that he had observed, that I wanted only a *good sort of hogs,* to make my stock complete. I thought, that I already had the finest in England; and I certainly had a very fine breed, the father of which, with legs not more than about six inches long, weighed, when he was killed, *twenty-seven score,* according to our Hampshire mode of stating hog-meat weight; or, *five hundred and forty pounds.* This breed has been fashioned by Mr. WOODS of Woodmancut in

Sussex, who has been, I believe, more than twenty years about it. I thought it perfection itself; but, I was obliged to confess, that Mr. GAUNTLETT's surpassed it.

102. Of the *earth burning* I will give an account in my next PART of this work. Nothing is easier of performance; and the materials are every where to be found.

103. I think, that I have now pretty clearly given an account of the modes of sowing, and planting, and cultivating the Ruta Baga, and of the preparation of the land. It remains for me to speak of the *time and manner of harvesting*, the *quantity of the crop*, and of the *uses* of, and the *mode of applying* the crop.

Time and Manner of Harvesting.

104. This must depend, in some measure, upon the *age* of the turnip; for, some will have their full growth earlier than others; that is to say, those, which are sown first, or transplanted first, will be *ripe* before those which are sown, or transplanted latest. I have made ample experiments as to this matter; and I will, as in former cases, first relate *what I did;* and then give my opinion as to what *ought* to be done.

105. This was a concern in which I could have no knowledge last fall, never having seen any turnips harvested in America, and knowing, that, as to American *frosts,* English expe-

rience was only likely to mislead; for, in England, we leave the roots standing in the ground all the winter, where we feed them off with sheep, which scoop them out to the very bottom; or we pull them as we want them, and bring them in to give to fatting oxen, to cows, or hogs. I had a great opinion of the *hardiness* of the Ruta Baga, and was resolved to *try* it here, and I did try it upon too large a scale.

106. I began with the piece, the first mentioned in paragraph 46: a part of them were taken up on the *13th of December*, after we had had some pretty hard frosts. The manner of doing the work was this. We took up the turnips merely by pulling them. The greens had been cut off and given to cattle before. It required a *spade* however, just to loosen them along the ridges, into which their tap-roots had descended very deeply. We dug holes at convenient distances, of a square form, and about a foot deep. We put into each hole about fifty bushels of turnips, piling them up above the level of the surface of the land, in a sort of pyramidical form. When the heap was made, we scattered over it about a truss of rye-straw, and threw earth over the whole to a thickness of about a foot, taking care to point the covering at top, in order to keep out wet.

107. Thus was a small part of the piece put up. The 14th of December was *a Sunday*, a

day that I can find no Gospel precept for devoting to the throwing away of the fruit of one's labours, and a day which I never will so devote again. However, I ought to have been *earlier*. On the Monday it *rained*. On the Monday night came a sharp North-Wester with its usual companion, at this season; that is to say, a *sharp frost*. Resolved to finish this piece on that day, I borrowed hands from my neighbours, who are always ready to assist one another. We had about two acres and a half to do; and it was necessary to employ about *one half* of the hands to go before the *pullers* and loosen the turnips with a spade in the frosty ground. About ten o'clock, I saw, that we should not finish, and there was every appearance of a hard frost at night. In order, therefore, to expedite the work, I called in the aid of those efficient fellow-labourers, a *pair of oxen*, which, with a good strong plough, going up *one side* of each row of turnips, took away the earth close to the bulbs, left them bare on one side, and thus made it extremely easy to pull them up. We wanted spades no longer; all our hands were employed taking up the turnips; and our job, instead of being half done that day, was completed by about *two o'clock*. Well and justly did MOSES order, that the ox should not be muzzled while he was treading out the corn; for, surely, no animals are so useful, so docile, so gentle as these, while they require at our hands so little care and labour in return!

108. Now, it will be observed, that the turnips here spoken of, were put up when the ground and the turnips *were frozen*. Yet they have kept perfectly sound and good; and I am preparing to plant some of them for seed. I am now writing on the *10th of April*. I send off these turnips to market every week. The tops and tails and offal to the pigs, to the ewes and lambs, and to a cow, and to working oxen, which all feed together upon this offal flung out about the barn-yard, or on the grass ground in the orchard. Before they have done, they leave not a morsel. But, of *feeding* I shall speak by and by.

109. The other crop of turnips, I mean those which were transplanted, as mentioned in paragraphs 72 and 73, and which, owing to their being planted so late in the summer, kept on *growing* most luxuriantly till the very hard frosts came.

110. We were now got on to the 17th of December; and I had *cabbages* to put up. Saturday, Sunday, and Monday, the 21st and 22nd and 23rd, we had a *very* hard frost, as the reader, if he live on this island, will well remember. There came a *thaw afterwards*, and the transplanted turnips were put up like the

others; but this hard frost had pierced them too deeply, especially as they were in so tender and luxuriant a state. Many of these we find rotted near the neck; and, upon the whole, they have suffered a loss of about *one half.* An acre, *left to take their chance in the field,* turned out, like most of the games of hazard, a *total loss.* They were *all* rotted.

111. This loss arose wholly from my want of sufficient experience. I was anxious to neglect no necessary precaution; and I was fully impressed, as I always am, with the advantages of being *early.* But, early in December, I lost a week at New York; and, though I worried my neighbours half to death to get at a knowledge of the time of the hard weather setting in, I could obtain no knowledge, on which I could rely, the several accounts being so different from each other. The general account was, that there would be no *very hard* weather till after Christmas. I shall know better another time! MAJOR CARTWRIGHT says, in speaking of the tricks of English Boroughmongers, at the "*Glorious* Revolution," that they will never be able to play the *same* tricks again; for that nations, like rational individuals, are not deceived *twice* in the same way.

112. Thus have I spoken of the *time and manner of harvesting,* as they took place with me. And, surely, the expense is a mere trifle. Two

oxen and four men would harvest two acres in any clear day in the latter end of November; and thus is this immense crop harvested, and covered completely, for about *two dollars* and a half an acre. It is astonishing, that this is never done in England! For, though it is generally said, that the Ruta Baga will stand *any* weather; I know, by experience, that it will not stand any weather. The winter of the year 1814, that is to say, the months of January and February, were very cold, and a great deal of snow fell; and, in a piece of twelve acres, I had, in the month of March, two thirds of the turnips *completely rotten;* and these were amongst the finest that I ever grew, many of them weighing twelve pounds each. Besides, when taken up in *dry weather*, before the freezings and thawings begin, the dirt all falls off; and the bulbs are clean and nice to be given to cattle or sheep in the stalls or yards. For, though we in general feed off these roots *upon the land* with sheep, we cannot, in deep land, always do it. The land is too *wet;* and particularly for ewes and lambs, which are, in such cases, brought into a piece of pasture land, or into a fold-yard, where the turnips are flung down to them in a *dirty* state, just carted from the field. And, again, the land is very much injured, and the labour augmented, by carting when the ground is a sort of mud-heap, or rather, pool. All these inconveniences and injuries would be avoided by harvesting in a dry day in November, if such a day should, by an accident, be found in England; but, why not do the work in October, and sow wheat, at once, in the land? More on this after-cropping, another time.

113. In Long Island, and throughout the United States, where the weather is so fine in the fall; where every day, from the middle of October to the end of November (except a rainy day about once in 16 days), is as fair as the fairest May-day in England, and where such a thing as a *water-furrow* in a field was never heard of; in such a soil as this, and under such a climate as this, there never can arise any difficulty in the way of the harvesting of turnips in proper time. I should certainly do it *in November;* for, as we have seen, a *little frost* does not affect the bulbs at all. I would put them in when perfectly *dry;* make my heaps of about fifty bushels; and, when the frosts approached, I mean the *hard* frosts, I would cover with corn-stalks, or straw, or cedar boughs, as many of the heaps as I thought I should want in January and February; for, these coverings would so break the frost, as to enable me to open the heaps in those severe months. It is useless and inconvenient to take into barns, or out-houses, a very large quantity at a time. Besides, if left *uncovered,* the very hard frosts

will do them harm. To be sure, this is easily prevented, in the barn, by throwing a little straw over the heap; but, being, by the means that I have pointed out, always kept ready in the field, to bring in a larger quantity than is used in *a week*, or thereabouts, would be wholly unnecessary, besides being troublesome from the great space, which would thus be occupied.

114. It is a great advantage in the cultivation of this crop, that the *sowing*, or transplanting time, comes *after* all the spring grain and the Indian Corn are safe in the ground, and *before* the harvest of grain begins; and then again, in the fall, the taking up of the roots comes after the grain and corn, and buck-wheat harvests, and even after the sowing of the winter grain. In short, it seems to me, that the cultivation of this crop, in this country, comes, as it were expressly, to fill up the unemployed spaces of the farmer's time; but, if he prefer standing with arms folded, during these spaces of time, and hearing his flock bleat themselves half to death in March and April, or have no flock, and scarcely any cattle or hogs, raise a few loads of yard-dung, and travel five miles for ashes, and buy them dear at the end of the five miles; if he *prefer* these, then, certainly, I shall have written on this subject in vain.

Quantity of the Crop.

115. It is impossible for me to say, at present, what quantity of Ruta Baga *may* be grown on an acre of land in this Island. My three acres of *ridged* turnips, sown on the 26th of June, were very unequal, but, upon one of the acres, there were *six hundred and forty bushels;* I mean *heaped bushels;* that is to say, an English statute bushel heaped as long as the commodity will lie on. The transplanted turnips yielded about *four hundred* bushels to the acre; but then, observe, they were put in a full month too late. This year, I shall make a fair trial.

116. I have given an account of my raising, upon five acres in one field, and twelve acres in another field, one thousand three hundred and twenty bushels to an acre, throughout the seventeen acres. I have no doubt of equalling that quantity on this Island, and that, too, upon some of its poorest and most exhausted land. They tell me, indeed, that the last summer was a *remarkably* fine summer; so they said at Botley, when I had my first prodigious crop of Ruta Baga. This is the case in all the pursuits of life. The moment a man excels those, who ought to be able and willing to do as well as he; that moment, others set to work to discover causes for his success, other than those proceeding *from himself*. But, as I used to tell

my neighbours at Botley, "You have had the "*same* seasons that I have had. Nothing is so "*impartial* as weather." As long as this sort of observation, or inquiry, proceeds from a spirit of *emulation*, it may be treated with great indulgence; but, when it discovers a spirit of *envy*, it becomes detestable, and especially in affairs of agriculture, where the appeal is made to our common parent, and where no man's success can be injurious to his neighbour, while it *must* be a benefit to his country, or the country in which the success takes place. I must, however, say, and I say it with feelings of great pleasure, as well as from a sense of justice, that I have observed in the American farmers *no envy* of the kind alluded to; but, on the contrary, the greatest *satisfaction*, at my success; and not the least backwardness, but great forwardness, to applaud and admire my mode of cultivating these crops. Not so, in England where the *farmers* (*generally* the most stupid as well as most slavish and most churlish part of the nation) envy all who excel them, while they are too obstinate to profit from the example of those whom they envy. I say *generally;* for there are many most honourable exceptions; and, it is amongst that class of men that I have my dearest and most esteemed friends; men of knowledge, of experience, of integrity, and of public-spirit, equal to that of the best of Englishmen in the worst times of oppression. I would not exchange the friendship of one of these men for that of all the Lords that ever were created, though there are some of them very able and upright men too.

117. Then, if I may be suffered to digress a little further here, there exists, in England, an institution, which has caused a sort of *identity of agriculture with politics.* The Board of Agriculture, established by Pitt for the purpose of sending *spies* about the country, under the guise of agricultural surveyors, in order to learn the cast of men's politics as well as the taxable capacities of their farms and property; this Board gives no premium or praise to any but "*loyal* farmers," who are generally the greatest fools. I, for my part, have never had any communication with it. It was always an object of ridicule and contempt with me; but, I know this to be the *rule* of that body, which is, in fact, only a little twig of the vast tree of corruption, which stunts, and blights, and blasts, all that approaches its poisoned purlieu. This Board has for its Secretary, Mr. ARTHUR YOUNG, a man of great talents, *bribed* from his good principles by this place of five hundred pounds a year. But Mr. YOUNG, though a most able man, is not always to be trusted. He is a bold asserter; and very few of his statements proceed upon actual experiments. And, as to what this

Board has *published,* at the public expense, under the name of *Communications,* I defy the world to match it as a mass of illiterate, unintelligible, useless trash. The only paper, published by this Board, that I ever thought worth keeping, was an account of the produce from a *single cow,* communicated by Mr. CRAMP, the jail-keeper of the County of Sussex; which contained very interesting and wonderful facts, properly authenticated, and stated in a clear manner.

118. ARTHUR YOUNG is blind, and never attends the Board. Indeed, sorrowful to relate, he is become a *religious fanatic,* and this in so desperate a degree as to leave no hope of any possible cure. In the pride of our health and strength, of mind as well as of body, we little dream of the chances and changes of old age. Who can read the "*Travels in France, Spain, and Italy,*" and reflect on the present state of the admirable writer's mind, without feeling some diffidence as to what may happen to himself!

119. LORD HARDWICKE, who is now the President of the Board, is a man, not exceeding my negro, either in experience or natural abilities. A parcel of court-sycophants are the Vice-Presidents. Their committees and correspondents are a set of justices of the peace, nabobs become country-gentlemen, and parsons of the

worst description. And thus is this a mere political job; a channel for the squandering of some thousands a year of the people's money upon worthless men, who ought to be working in the fields, or mending "His Majesty's Highways."

120. Happily, politics, in this country, have nothing to do with agriculture; and here, therefore, I think I have a chance to be fairly heard. I should, indeed, have been heard in England; but, I really could never bring myself to do any thing tending to improve the estates of the oppressors of my country; and the same consideration now restrains me from communicating information, on the subject of timber trees, which would be of immense benefit to England; and which information I shall reserve, till the tyranny shall be at an end. Castlereagh, in the fulness of his stupidity, proposed, that, in order to find employment for "*the population*," as he insolently called *the people of England*, he would set them to dig holes one day and fill them up the next. I could tell him what to *plant* in the holes, so as to benefit the country in an immense degree; but, like the human body in some complaints, the nation would now be really injured by the communications of what, if it were in a healthy state, would do it good, add to its strength, and to all its means of exertion.

121. To return from this digression, I am afraid of *no bad seasons.* The *drought,* which is the great enemy to be dreaded in this country, I am quite prepared for. Give me ground that I can plough ten or twelve inches deep, and give me Indian corn spaces to plough in, and no sun can burn me up. I have mentioned Mr. CURWEN's experiment before; or, rather TULL's; for he it is, who made all the discoveries of this kind. Let any man, just to try, leave half a rod of ground *undug* from the month of May to that of October; and another half rod to let him *dig* and *break fine* every ten or fifteen days. Then, whenever there has been fifteen days of good scorching sun, let him go and dig a hole in each. If he does not find the hard ground *dry as dust,* and the other *moist;* then let him say, that I know nothing about these matters. So erroneous is the common notion, that ploughing *in dry weather lets in the drought!*

122. Of course, proceeding upon this fact, which I state as the result of numerous experiments, I should, if visited with long droughts, give one or two additional ploughings between the crops when growing. That is all; and, with this, in Long Island, I defy all droughts.

123. But, why need I insist upon this effect of ploughing in dry weather? Why need I insist on it in an Indian corn country? Who has not seen fields of Indian corn looking, to-day,

yellow and sickly, and, in four days hence (the weather being dry all the while), looking green and flourishing; and this wonderful effect produced merely by the plough? Why, then, should not the same effect always proceed from the same cause? The *deeper* you plough, the greater the effect, however; for there is a greater body of earth to exhale from, and to receive back the tribute of the atmosphere. Mr. CURWEN tells us of a piece of cattle-cabbage. In a very dry time in July, they looked so *yellow* and *blue*, that he almost despaired of them. He sent in his ploughs; and a gentleman, who had seen them when the ploughs went in on the Monday, could scarcely believe his eyes when he saw them on the next Saturday, though it had continued dry all the week.

124. To perform these summer ploughings, in this Island, is really nothing. The earth is so light and in such fine order, and so easily displaced and replaced. I used one horse for the purpose, last summer, and a very slight horse indeed. *An ox* is, however, better for this work; and this may be accomplished by the use of a *collar* and two traces, or by a *single yoke* and two traces. TULL recommends the latter; and I shall try it for Indian corn as well as for turnips.* Horses, if they are strong

* Since the above paragraph was written, I have made a *single ox-yoke*; and, I find it answer excellently well. Now, my work is much shortened; for, in forming ridges, two oxen

enough, are not so *steady* as oxen, which are more patient also, and with which you may send the plough-share *down* without any of the

are *awkward.* They occupy a wide space, and one of them is obliged to walk upon the ploughed land, which, besides making the ridge uneven at top, presses the ground, which is injurious. For ploughing between the rows of turnips and *Indian corn* also, what a great convenience this will be! An ox goes *steadier* than a horse, and will plough *deeper,* without fretting and without tearing; and he wants neither *harness-maker* nor *groom.* The plan of my yoke I took from TULL. I showed it to my workman, who chopped off the limb of a tree, and made the yoke in an hour. It is a piece of wood, with two holes to receive two ropes, about three quarters of an inch in diameter. These traces are fastened into the yoke merely by a knot, which prevents the ends from passing through the holes, while the other ends are fastened to the two ends of a *Wiffle-tree,* as it is called in Long Island, of a *Wipple-tree* as it is called in Kent, and of a *Wippauce,* as it is called in Hampshire. I am but a poor draftsman; but, if the printer can find any thing to make the representation with, the following draft will clearly show what I have meant to describe in words—

When the corn (Indian) and turnips get to a size, sufficient to attract the *appetite* of the ox, you have only to put on a *muzzle.* This is what Mr. TULL did; for, though we ought not to muzzle the ox "as he *treadeth out the corn,*" we may do it, even for his own sake, amongst other considerations, when he is assisting us to bring the crop to perfection.

fretting and unequal pulling, or jerking, that you have to encounter with horses. And, as to the *slow pace* of the ox, it is the old story of the tortoise and the hare. If I had known, in England, of the use of oxen, what I have been taught upon Long Island, I might have saved myself some hundreds of pounds a year. I ought to have followed TULL in this as in all other parts of his manner of cultivating land. But, in our country, it is difficult to get a ploughman to look at an ox. In this Island the thing is done so completely and so easily, that it was, to me, quite wonderful to behold. To see one of these Long-Islanders going into the field, or orchard, at sun-rise, with his yoke in his hand, call his oxen by name to come and put their necks under the yoke, drive them before him to the plough, just hitch a hook on to the ring of the yoke, and, then, without any thing except a single chain and the yoke, with no reins, no halter, no traces, no bridle, no driver, set to plough, and plough a good acre and a half in the day, To see this would make an English farmer stare; and well it might, when he looked back to the ceremonious and expensive business of keeping and managing a plough-team in England.

125. These are the means, which I would, and which I shall, use to protect my crops against the effects of a *dry season.* So that, as

every one has the same means at his command, no one need be afraid of drought. It is a *bright* plough-share that is always wanted much more than the showers. With this culture there is no fear of a crop; and though it amount to only five hundred bushels on an acre, what crop is *half* so valuable.

126. The *bulk of crop,* however, in the *broad-cast,* or random, method, may be materially affected by *drought;* for in that case, the plough cannot come to supply the place of showers. The ground there will be dry, and keep dry in a dry time; as in the case of the supposed half rod of undug ground in the garden. The weeds, too, will come and help, by their roots, to suck the moisture out of the ground. As to the *hand-hoeings,* they may keep down weeds to be sure, and they raise a trifling portion of exhalation; but, it is trifling indeed. Dry weather, if of long continuation, makes the leaves become of a *bluish* colour; and, when this is once the case, all the rain and all the fine weather in the world will never make the crop a good one; because the plough cannot move amidst this scene of endless irregularity. This is one of the chief reasons why the ridge method is best.

Uses of, and Mode of applying, the Crop.

127. It is harder to say what uses this root may not be put to, than what uses it may

be put to, in the feeding of animals. It is eaten greedily by sheep, horn-cattle, and hogs, in its raw state. *Boiled*, or *steamed* (which is better), no *dog* that I ever saw, will refuse it. Poultry of all sorts will live upon it in its cooked state. Some dogs will even eat it raw; a fact that I first became acquainted with by perceiving my Shepherd's dog eating in the field along with the sheep. I have two *Spaniels* that come into the barn and eat it now; and yet they are both in fine condition. Some horses will nearly live upon it in the raw state; others are not so fond of it.

128. Let me give an account of what I am doing now (in the month of April) with my crop.

129. It is not pretended, that this root, *measure for measure*, is equal to *Indian corn in the ear*. Therefore, as I can get Indian corn in the ear for half a dollar a bushel, and, as I sell my Ruta Baga for half a dollar a bushel at New York, I am very sparing of the use of the latter for animals. Indeed, I use none at home, except such as have been injured, as above-mentioned, by the delay in the harvesting. These damaged roots I apply in the following manner.

130. Twice a day I take about two bushels, and scatter them about upon the grass for fifteen ewes with their lambs, and a few wether sheep, and for seven stout store pigs, which eat with

them. Once a day I fling out a parcel of the refuse that have been cut from the roots sent to market, along with cabbage leaves and stems, parsnip fibres, and the like. Here the working oxen, hogs, cows, sheep, and fowls, all feed as they please. All these animals are in excellent condition. The cow has *no other* food; the working oxen a lock of hay twice a day; the ewes an ear of Indian corn each; the pigs nothing but the roots; the fowls and ducks and turkeys are never *fed* in any other way, though they know how to feed themselves whenever there is any thing good to be found above ground.

131. I am *weaning* some pigs, which, as every one knows, is an affair of *milk* and *meal.* I have neither. I give about three buckets of *boiled* Ruta Baga to *seven pigs* every day, not having any convenience for steaming; two *baits* of *Indian corn in the ear.* And, with this diet, increasing the quantity with the growth of the pigs, I expect to turn them out of the sty fatter (if that be possible) than they entered it. Now, if this *be so,* every farmer will say, that this is what never was done before in America. We all know how important a thing it is to *wean* a pig *well.* Any body can wean them without *milk* and *meal;* but, then, the pigs are good for nothing. They remain three months afterwards and never grow an inch; and they

are, indeed, not worth having. To have milk, you must have *cows*, and cows are vast consumers! To have cows, you must have *female labour*, which, in America, is a very precious commodity. You cannot have *meal* without sharing in kind pretty liberally with the miller, besides bestowing labour, however busy you may be, to carry the corn to mill and bring the meal back. I am, however, speaking here of the pigs from my English breed; though I am far from supposing that the common pigs might not be weaned in the same way.

132. *Sows with young pigs* I feed thus: boiled Ruta Baga twice a day. About three ears of Indian corn a piece twice a day. As much offal Ruta Baga *raw* as they will eat. Amongst this boiled Ruta Baga, the pot-liquor of the house goes, of course; but, then, the dogs, I dare say, take care that the best shall fall to their lot! and as there are four of them pretty fat, their share cannot be very small. Every one knows, what good food, how much *meal* and *milk* are necessary to sows which have pigs. I have no milk, for my cow has not yet calved. And, then, what a *chance* concern this is; for, the sows may perversely have pigs at the time when the cows *do not please to give milk;* or, rather, when they, poor things, without any fault of theirs, are permitted to *go dry*, which never need be, and never

ought to be the case. I had a cow once that made more than two pounds of butter during the week, and had a calf on the Saturday night. Cows always ought to be milked to the very day of their calving, and during the whole of the time of their suckling their calves. But, "sufficient unto the day is the evil thereof." Let us leave this matter till another time. Having, however, accidentally mentioned *cows*, I will just observe, that in the little publication of Mr. CRAMP, mentioned above, as having been printed by the Board of Agriculture, it was stated and the proof given, that his *single cow* gave him, *clear profit*, for several successive years, more than *fifty pounds sterling a year*, or upwards of *two hundred and twenty dollars*. This was *clear profit;* reckoning the food and labour, and taking credit for the calf, the butter, and for the skim-milk at a *penny a quart* only. Mr. CRAMP's was a *Sussex* cow. Mine were of the *Alderney* breed. Little, small-boned things; but, two of my cows, fed upon *three quarters of an acre of grass ground*, in the middle of my shrubbery, and fastened to pins in the ground, which were shifted twice a day, made *three hundred pounds of butter* from the 28th of March to the 27th of June. This is a finer country for cattle than England; and yet, what do I see!

133. This difficulty about feeding sows with

young pigs and weaning pigs, is one of the greatest hindrances to improvement; for, after all, what animal produces flesh meat like the hog? applicable to all uses, either fresh or salted, is the meat. Good in all its various shapes. The animal killable at all ages. Quickly fatted. Good if half fat. Capable of supporting an immense burden of fat. Demanding but little space for its accommodation; and yet, if grain and corn and milk are to be their principal food, during their lives, they cannot multiply very fast; because many upon a farm cannot be kept to much profit. But, if by providing a sufficiency of Ruta Baga, a hundred pigs could be raised upon a farm in a year, and carried on till fatting time, they would be worth, when ready to go into the fatting sty, fifteen dollars each. This would be something worth attending to; and the farm *must* become rich from the manure. The Ruta Baga, taken out of the heaps early in April, will keep well and sound all the summer; and with a run in an orchard, or in a grassy place, it will keep a good sort of hog always in a very thriving, and even *fleshy* state.

134. This root, being called a *turnip*, is regarded as a *turnip*, as a common turnip, than which nothing can be much less resembling it. The common turnip is a very poor thing. The poorest of all the roots of the bulb kind, cul-

tivated in the fields; and the Ruta Baga, all taken together, is, perhaps, the very best. It loses none of its good qualities by being long kept, though dry all the while. A neighbour of mine in Hampshire, having saved a large piece of Ruta Baga for *seed*, and having, after harvesting the seed, accidentally thrown some of the roots into his yard, saw his hogs eat these old roots, which had borne the seed. He gave them some more, and saw that they ate them greedily. He, therefore, went and bought a whole drove, in number about forty, of lean pigs, of a good large size, brought them into his yard, carted in the roots of his seed Ruta Baga, and, without having given the pigs a handful of any other sort of food, sold out his pigs as *fat porkers*. And, indeed, it is a fact well known, that sheep and cattle, as well as hogs, will thrive upon this root after it has borne seed, which is what, I believe, can be said of no other root or plant.

135. When we feed off our Ruta Baga in the fields, in England, by sheep, there are small parts left by the sheep: the *shells* which they have left after scooping out the pulp of the bulb; the tap-root; and other little bits. These are picked out of the ground; and when washed by the rain, other sheep follow and live upon them. Or, in default of other sheep, hogs or cattle are turned in in dry weather, and they leave not a morsel.

136. Nor are the *greens* to be forgotten. In England, they are generally eaten by the sheep, when they are turned in upon them. When the roots are taken up for uses at the home-stead, the greens are given to store-pigs and lean cattle. I cut mine off, while the roots were in the ground, and gave them to fatting cattle upon grass land, alternately with Indian corn in the ear; and, in this way, they are easily and most profitably applied, and they come, too, just after the grass is gone from the pastures. An acre produces about four good waggon loads of greens; and they are taken off fresh and fresh as they are wanted, and, at the same time, the roots are thus made ready for going, at once, into the heaps. Pigs, sheep, cattle; all like the greens as well as they do the roots. Try any of them with the greens of *white* turnips; and, if they touch them, they will have changed their natures, or, at least, their tastes.

137. The Mangel Wurzel, the cabbage, the carrot, and the parsnip, are all useful; and the *latter*, that is to say, the *parsnip*, very valuable indeed; but the *main* cattle-crop is the Ruta Baga. Even the *white* turnip, if well cultivated, may be of great use; and, as it admits of being *sown later*, it may often be very desirable to raise it. But, reserving myself to speak fully, in a future part of my work, of my experiments as to these crops, I shall now make

a short inquiry as to the value of a crop of Ruta Baga, compared with the value of any other crop. I will just observe, in this place, however, that I have grown *finer* carrots, parsnips, and Mangel Wurzel, and even finer cabbages, than I ever grew upon the richest land in Hampshire, though not a seed of any of them was put into the ground till the *month of June*.

138. A good mode, it appears to me, of making my proposed comparative estimate, will be to say, *how I would proceed*, supposing me to have a farm of my own in this island, of only one hundred acres. If there were not twelve acres of orchard, near the house, I would throw as much grass land to the orchard as would make up the twelve acres, which I could fence in an effectual manner, against small pigs as well as large oxen.

139. Having done this, I would take care to have fifteen acres of *good* Indian corn, well planted, well suckered, and well tilled in all respects. Good, deep ploughing between the plants would give me forty bushels of shelled corn to an acre; and a ton to the acre of fodder for my four working oxen and three cows, and my sheep and hogs, of which I shall speak presently.

140. I would have *twelve* acres of Ruta Baga, *three* acres of early cabbages, an acre of Mangel

Wurzel, an acre of carrots and parsnips, and as many white turnips as would grow between my rows of Indian corn after my last ploughing of that crop.

141. With these crops, which would occupy thirty-two acres of ground, I should not fear being able to keep a good house in all sorts of meat, together with butter and milk, and to send to market nine quarters of beef and three hides, a hundred early fat lambs, a hundred hogs, weighing twelve score, as we call it in Hampshire, or, two hundred and forty pounds each, and a hundred fat ewes. These, altogether, would amount to about three thousand dollars, exclusive of the cost of a hundred ewes and of three oxen; I should hope, that the produce of my trees in the orchard and of the other fifty-six acres of my farm would pay the rent and the labour; for, as to *taxes*, the amount is not worth naming, especially after the sublime spectacle of that sort, which the world beholds in England.

142. I am, you will perceive, not making any account of the price of Ruta Baga, cabbages, carrots, parsnips, and white turnips at *New York*, or any other market. I *now*, indeed, sell carrots and parsnips at three quarters of a dollar *the hundred*, by tale; cabbages (of last fall) at about three dollars a hundred, and white turnips at a quarter of a dollar a bushel. When this can

be done, and the distance is within twenty or thirty miles on the *best road in the world*, it will, of course, be done; but, my calculations are built upon a supposed consumption of the whole upon the farm by animals of one sort or another.

143. My feeding would be nearly as follows. I will begin with February; for, until then, the Ruta Baga does not come to its sweetest taste. It is like an apple, that must have time to ripen; but, then, it retains its goodness much longer. I have proved, and especially in the feeding of hogs, that the Ruta Baga is never so good, till it arrives at a mature state. In February, and about the first of that month, I should begin bringing in my Ruta Baga, in the manner before described. My three oxen, which would have been brought forward by other food to be spoken of by and by, would be *tied* up in a stall looking into one of those fine commodious barn's floors which we have upon this island. Their stall should be *warm*, and they should be kept well littered, and cleaned out frequently. The Ruta Baga just chopped into large pieces with a spade or shovel, and tossed into the manger to the oxen at the rate of about two bushels a day to each ox, would make them completely fat, without the aid of corn, hay, or any other thing. I should, probably, kill one ox at Christmas, and, in that case,

he must have had a longer time than the others upon other food. If I killed one of the two remaining oxen in the middle of March, and the other on the first of May, they would consume 266 bushels of Ruta Baga.

144. My hundred ewes would begin upon Ruta Baga at the same time, and, as my grass ground would be only twelve acres until after hay-time, I shall suppose them to be fed on this root till July, and they will always eat it and thrive upon it. They will eat about eight pounds each, a day; so that, for 150 days it would require a hundred and twenty thousand pounds weight, or two thousand four hundred bushels.

145. Fourteen breeding sows to be kept all the year round, would bring a hundred pigs in the Spring, and they and their pigs would, during the same 150 days, consume much about the same quantity; for, though the pigs would be small during these 150 days, yet they eat a great deal more than sheep in proportion to their size, or rather bulk. However, as they would eat very little during 60 days of their age, I have rather over-rated their consumption.

146. Three cows and four working oxen would, during the 150 days consume about one thousand bushels, which, indeed, would be more than sufficient, because, during a great part of the time, they would more than half live

upon corn-stalks; and indeed this, to a certain extent, would be the case with the sheep. However, as I mean that every thing should be of a good size, and *live well*, I make ample provision.

147. I should want then, to raise *five hundred* bushels of Ruta Baga upon each of my twelve acres; and why should I not do it, seeing that I have this year raised *six hundred and forty* bushels upon an acre, under circumstances such as I have stated them? I lay it down, therefore, that, with a culture as good as that of Indian corn, any man may, on this island (where corn will grow) have 500 bushels to the acre.

148. I am now come to the first of July. My oxen are fatted and disposed of. My lambs are gone to market, the last of them a month ago. My pigs are weaned and of a good size. And now my Ruta Baga is gone. But my ewes, kept well through the winter, will soon be fat upon the 12 acres of orchard and the hay-ground, aided by my three acres of early cabbages, which are now fit to begin cutting, or, rather, pulling up. The weight of this crop may be made very great indeed. Ten thousand plants will stand upon an acre, in *four feet ridges*, and every plant ought to weigh *three pounds* at least. I have shown before how advantageously Ruta Baga *transplanted* would follow these cabbages, all through the months of July and August. But

what a crop of *Buck-wheat* would follow such of the cabbages as came off in *July!* My cabbages, together with my hay-fields and grain-fields after harvest, and about forty or fifty waggon-loads of Ruta Baga *greens*, would carry me along well till December (the cabbages being planted at different times); for my ewes would be sold fat in July, and my pigs would be only increasing in demand for food; and the new hundred ewes need not, and ought not, to be kept so well as if they were fatting, or had lambs by their side.

149. From the first of December to the first of February, Mangel Wurzel and white turnips would keep the sheep and cattle and breeding sows plentifully; for the latter will live well upon Mangel Wurzel; and my hundred hogs, intended for fatting, would be much *more* than *half* fat upon the carrots and parsnips. I should, however, more probably keep my parsnips till Spring, and mix the feeding with carrots with the feeding with corn, for the first month, or fifteen days, with regard to the fatting hogs. None of these hogs would require more than three bushels of corn each to finish them completely. My other three hundred bushels would be for sows giving suck; the ewes, now and then in wet weather; and for other occasional purposes.

150. Thus all my *hay* and *oats*, and *wheat* and

rye might be sold, leaving me the straw for litter. These, surely, would pay the rent and the labour; and, if I am told, that I have taken no account of the mutton, and lamb, and pork, that my house would demand, neither have I taken any account of *a hundred summer pigs*, which the fourteen sows would have, and which would hardly fail to bring two hundred dollars. Poultry demand some *food;* but three parts of their raising consists of *care;* and, if I had nobody in my house to bestow this care, I should, of course, have the less number of mouths to feed.

151. But, my *horses!* Will not they swallow my *hay* and my *oats?* No: for I want no horses. But, am I never to *take a ride*, then? Aye, but, if I do, I have no right to lay the expense of it to the account of the *farm*. I am speaking of how a man may live by and upon *a farm*. If a merchant spend a thousand a year, and gain a thousand, does he say, that his traffic has gained him nothing? When men *lose money by farming*, as they call it, they forget, that it is not *the farming*, but other expenses that take away their money. It is, in fact, they that rob the farm, and not the farm them. Horses may be kept for the purposes of going to church, or to meeting, or to pay visits. In many cases this may be not only convenient, but necessary, to a family; but, upon this Island, I am very sure,

that it is neither convenient nor necessary to *a farm.* "What!" the ladies will say, "would "you have us to be shut up at home all our "lives; or be dragged about by oxen?" By no means; not I! I should be very sorry to be thought the author of any such advice. I have no sort of objection to the keeping of horses upon a farm; but, I do insist upon it, that all the food and manual labour required by such horses, ought to be considered as so much taken from the clear profits of the farm.

152. I have made sheep, and particularly *lambs,* a part of my supposed stock; but, I do not know, that I should keep any beyond what might be useful for my house. ***Hogs*** are the most profitable stock, if you have a large quantity of the food that they will *thrive* on. They are *foul* feeders; but, they will eat nothing that is poor in its nature; that is to say, they will not *thrive* on it. They are the most able *tasters* in all the creation; and, that which they like best, you may be quite sure has the greatest proportion of nutritious matter in it, from a white turnip to a piece of beef. They will prefer meat to corn, and cooked meat to raw; they will leave parsnips for corn or grain; they will leave carrots for parsnips; they will leave Ruta Baga for carrots; they will leave cabbages for Ruta Baga; they will leave Mangel Wur-

zel for cabbages; they will leave potatoes (both being raw) for Mangel Wurzel. A *white* turnip they will not touch, unless they be on the point of starving. They are the best of *triers*. Whatever they prefer is sure to be the *richest* thing within their reach. The parsnip is, by many degrees, the richest root; but, the seed lies long in the ground; the sowing and after-culture are works of great niceness. The crop is large with good cultivation; but, as a *main* crop, I prefer the Ruta Baga, of which the crop is immense, and the harvesting, and preserving, and application of which, are so easy.

153. The farm I suppose to be in *fair condition* to start with: the usual grass-seeds sown, and so forth; and every farmer will see, that, under my system, it must soon become rich as any garden need to be, without my sending men and horses to the water-side to fetch ashes, which have been brought from Boston or Charleston, an average distance of seven hundred miles! In short, my stock would give me, in one shape or another, manure to the amount, in utility, of more than a thousand tons weight a year of common yard manure. This would be ten tons to an acre *every year*. The farm would, in this way, become more and more productive; and, as to its being *too rich*, I see no danger of that; for a broad-cast crop

of wheat will, at any time, tame it pretty sufficiently.

154. Very much, in my opinion, do those mistake the matter, who strive to get a *great breadth* of land, with the idea, that, when they have *tried* one field, they can let it lie, and go to another. It is better to have one acre of good crop, than two of bad or indifferent. If the one acre can by *double* the manure and *double* the labour in tillage, be made to produce as much as two other acres, the one acre is preferable, because it requires only half as much fencing and little more than half as much harvesting, as two acres. There is many a ten acres of land near London, that produce more than any common farm of two hundred acres. My garden of *three quarters of an acre*, produced more, in value, last Summer, from June to December, than any *ten acres* of oat land upon Long Island, though I there saw as fine fields of oats as I ever saw in my life. A *heavy crop* upon *all* the ground that I put a plough into is what I should seek, rather than to have a great quantity of land.

155. The business of carting manure from a *distance* can, in very few, if any cases, answer a profitable purpose. If any man would give me even horse-dung at the stable-door, four miles from my land, I would not accept of it,

on condition of fetching it. I say the same of spent ashes. To manure a field of ten acres, in this way, a man and two horses must be employed twenty days at least, with twenty days wear and tear of waggon and tackle. Two oxen and two men do the business in two days, if the manure be on the spot.

156. In concluding my remarks on the subject of Ruta Baga, I have to apologize for the desultory manner in which I have treated the matter; but, I have put the thoughts down as they occurred to me, without much time for arrangement, wishing very much to get this first Part into the hands of the public before the arrival of the time for sowing Ruta Baga this present year. In the succeeding Parts of the work, I propose to treat of the culture of every other plant that I have found to be of use upon a farm; and also to speak fully of the *sorts* of cattle, sheep, and hogs, particularly the latter. My experiments are now going on; and, I shall only have to communicate the result, which I shall do very faithfully, and with as much clearness as I am able. In the mean while, I shall be glad to afford any opportunity, to any persons who may think it worth while to come to Hyde Park, of *seeing* how I proceed. I have just now (17th April) planted out my Ruta Baga, Cabbages, Mangel Wurzel, Onions, Parsnips, &c. for seed. I shall begin my *earth-*

burning in about fifteen days. In short, being convinced, that I am able to communicate very valuable improvements; and not knowing how short, or how long, my stay in America may be, I wish very much to leave behind me whatever of good I am able, in return for the protection, which America has afforded me against the fangs of the Boroughmongers of England; to which country, however, I always bear affection, which I cannot feel towards any other in the same degree, and the prosperity and honour of which I shall, I hope, never cease to prefer before the gratification of all private pleasures and emoluments.

END

Of the Treatise on Ruta Baga,

AND OF PART I.

A

YEAR'S RESIDENCE,

IN THE

UNITED STATES OF AMERICA.

Treating of the Face of the Country, the Climate, the Soil, the Products, the Mode of Cultivating the Land, the Prices of Land, of Labour, of Food, of Raiment; of the Expenses of Housekeeping, and of the usual manner of Living; of the Manners, Customs, and Character of the People; and of the Government, Laws, and Religion.

IN THREE PARTS.

By WILLIAM COBBETT.

PART II.

Containing,—III. Experiments as to Cabbages.—IV. Earth-burning.—V. Transplanting Indian Corn.—VI. Swedish Turnips.—VII. Potatoes.—VIII. Cows, Sheep, Hogs, and Poultry.—IX. Prices of Land, Labour, Working Cattle, Husbandry Implements.—X. Expenses of Housekeeping.—XI. Manners, Customs, and Character of the People.—XII. Rural Sports.—XIII. Paupers and Beggars.—XIV. Government, Laws, and Religion.

LONDON:

PRINTED FOR SHERWOOD, NEELY AND JONES,

PATERNOSTER-ROW.

1819.

CONTENTS OF PART II.

DEDICATION

TO

Mr. RICHARD HINXMAN

OF CHILLING IN HAMPSHIRE.

North Hempstead, Long Island,
15th Nov. 1818.

MY DEAR SIR,

THE following little volume will give you some account of my agricultural proceedings in this fine and well-governed country; and, it will also enable you to see clearly how favourable an absence of grinding taxation and tithes is to the farmer. You have already paid to Fundholders, Standing Armies, and Priests, more money than would make a decent fortune for two children; and, if the present system were to continue to the end of your natural life, you would pay more to support the idle and the worthless, than would maintain, during the same space of time, ten labourers and their families. The profits of your capital, care and skill are pawned by the Boroughmongers to pay the interest of a Debt, which they have contracted for their own purposes; a Debt, which never can, by ages of toil and of sufferings, on the part of the people, be either paid off or diminished. But, I trust, that deliverance from this worse than Egyptian bondage is now near at hand. The atrocious tyranny does but stagger along. At every step it discovers fresh proofs of impotence. It must come down; and when it is down, we shall not have to envy the farmers of America, or of any country in the world.

When you reflect on the blackguard conduct of the *Parsons* at Winchester, on the day when I last had the pleasure to see you and our excellent friend Goldsmith, you will rejoice to find, that, throughout the whole of this extensive country, there exists not one single animal of that description; so that we can here keep as many cows, sows, ewes and hens as we please, with the certainty, that no prying, greedy Parson will come to eat up a part of the young ones. How long shall we Englishmen suffer our cow-stalls, our styes, our folds and our hen-roosts to be the prey of this prowling pest?

In many parts of the following pages you will trace the remarks and opinions back to conversations that have passed between us, many times in Hampshire. In the making of them my mind has been brought back to the feelings of those days. The certainty, that I shall always be beloved by you constitutes one of the greatest pleasures of my life; and I am sure, that you want nothing to convince you, that I am unchangeably

Your faithful and affectionate friend,

WM. COBBETT.

PREFACE

TO THE

SECOND PART.

---

157. In the First Part I adopted the mode of *numbering the paragraphs,* a mode which I shall pursue to the end of the work; and, as the whole work may, at the choice of the purchaser, be bound up in one volume, or remain in two volumes, I have thought it best to resume the numbering at the point where I stopped at the close of the First Part. The last paragraph of that Part was 156: I, therefore, now begin with 157. For the same reason I have, in the Second Part, resumed the *paging* at the point where I stopped in the First Part. I left off at page 186; and, I begin with 187. I have, in like manner, resumed the *chaptering:* so that, when the two volumes are put together, they will, as to these matters, form but one; and those, who may have purchased the volumes separately, will possess the same book, in all respects, as those, who shall purchase the Three Parts in one Volume.

158. Paragraph 1. (Part I.) contains my reasons for numbering the paragraphs, but, besides the reasons there stated, there is one, which did not then occur to me, and which was left to be suggested by experience, of a description which I did not then anticipate; namely, that, in the case of *more than one edition*, the *paging* may, and generally does, differ in such manner as to bring the matter, which, in one edition, is under any given page, under a different page in another edition. This renders the work of *reference* very laborious at best, and, in many cases, it defeats its object. If the paragraphs of BLACKSTONE'S COMMENTARIES had been numbered, how much valuable time it would have saved. I am now about to send a second edition of the First Part of this work to the press. I am quite careless about the *paging:* that is to say, so that the whole be comprized within the 134 pages, it is of no consequence whether the matter take, with respect to the pages, precisely the same situation that it took before; and, if the paging were not intended to join on to that of the present volume, it would be no matter what were the number of pages upon the whole. I hope, that these reasons will be sufficient to convince the reader that I have not, in this case, been actuated by a love of sin-

gularity. We live to learn, and to make improvements, and every improvement must, at first, be a singularity.

159. The utility, which I thought would arise from the *hastening* out of the First Part, in June last, previous to the time for sowing Swedish Turnips, induced me to make an ugly breach in the *order* of my little work; and, as it generally happens, that when disorder is once begun, it is very difficult to restore order; so, in this case, I have been exceedingly puzzled to give to the matter of these two last Parts such an arrangement as should be worthy of a work, which, whatever may be the character of its execution, treats of subjects of great public interest. However, with the help of the Index, which I shall subjoin to the Third Part, and which will comprise a reference to the divers matters in all the three parts, and in the making of which Index an additional proof of the advantage of numbering the paragraphs has appeared; with the help of this Index the reader will, I am in hopes, be enabled to overcome, without any very great trouble, the inconveniences naturally arising from a want of a perfectly good arrangement of the subjects of the work.

160. As the First Part closes with a promise to communicate the result of my experiments of this present year, I begin the *Second*

Part with a fulfilment of that promise, particularly with regard to the *procuring of manure by the burning of earth into ashes.*

161. I then proceed with the other matters named in the title; and the *Third Part* I shall make to consist of an account of the *Western Countries,* furnished in the Notes of Mr. HULME, together with a view of the advantages and disadvantages of preferring, as a place to farm in, those Countries to the Countries bordering on the Atlantic; in which view I shall include such remarks as appear to me likely to be useful to those *English Farmers,* who can no longer bear the lash of Boroughmongering oppression and insolence.

162. Multifariousness is a great fault in a written work of any kind. I feel the consciousness of this fault upon this occasion. The facts and opinions relative to Swedish Turnips and Cabbages will be very apt to be enfeebled in their effect by those relating to manners, laws and religion. Matters so heterogeneous, the one class treated of in the detail and the other in the great, ought not to be squeezed together between the boards of the same small volume. But, the fault is committed and it is too late to repine. There are, however, two subjects which I will treat of distinctly hereafter. The first is that of *Fencing,* a subject

which presses itself upon the attention of the American Farmer, but from which he turns with feelings like those, with which a losing tradesman turns from an examination of his books. But, attend to it he *must* before it be long; or, his fields, in the populous parts of this Island at least, must lay waste, and his fuel must be brought him from Virginia or from England. Sometime before March next I shall publish an *Essay on Fencing*. The form shall correspond with that of this work, in order that it may be bound up with it, if that should be thought desirable. The other subject is that of *Gardening*. This I propose to treat of in a small distinct volume, under some appropriate title; and, in this volume, to give *alphabetically*, a description of *all the plants*, cultivated for the use of the *table* and also of those cultivated as *cattle food*. To this description I shall add an account of their properties, and instructions for the cultivation of them in the best manner. It is not my intention to go beyond what is aptly enough called the *Kitchen Garden;* but, as a *hot-bed* may be of such great use even to the farmer; and as ample materials for making beds of this sort are *always* at *his* command without any *expence*, I shall endeavour to give plain directions for the making and managing of a hot-bed. A bed of this sort, fifteen feet long, has given me, this year, the better part of an acre of fine cabbages to give to hogs in the parching month of *July*. This is so very simple a matter; it is so very easy to learn; that there is scarcely a farmer in America, who would not put the thing in practice, at once, with complete success.

163. Let not my countrymen, who may happen to read this suppose, that these, or any other, pursuits will withdraw my attention from, or slacken my zeal in, that cause, which is common to us all. That cause claims, and has, my first attention and best exertion; that is the *business* of my life: these other pursuits are my *recreation*. King ALFRED allowed eight hours for *recreation*, in the twenty-four, eight for *sleep*, and eight for *business*. I do not take my allowance of the two former.

164. Upon looking into the First Part, I see, that I expressed a hope to be able to give, in some part of this work, a sketch of the work of Mr. TULL. I have looked at TULL, and I cannot bring my mind up to the commission of so horrid an act as that of garbling such a work. It was, perhaps, a feeling, such as that which I experience at this moment, which restrained Mr. CURWEN from even *naming* TULL, when he gave one of TULL's experiments to the

world as a *discovery of his own.* Unable to screw himself up to commit a murder, he contented himself with a robbery; an instance, he may, indeed, say, of singular moderation and self-denial; especially when we consider of what an assembly he has, with little intermission, been an " Honourable Member" for the last thirty years of his life.

WM. COBBETT.

North Hempstead, Long Island,
15*th November,* 1818.

A

YEAR'S RESIDENCE,

&c.

CHAP. III.

EXPERIMENTS, IN 1818, AS TO CABBAGES.

Preliminary Remarks.

165. At the time when I was writing the First Part, I expected to be able to devote more time to my farming, during the summer, than I afterwards found that I could so devote without neglecting matters which I deem of greater importance. I was, indeed, obliged to leave the greater part of my out-door's business wholly to my men, merely telling them what to do. However, I attended to the things which I thought to be of the most importance. The field-culture of Carrots, Parsnips and Mangle Wurzle I did not attempt. I contented myself with a crop of Cabbages and of Ruta Baga and with experiments as to Earth-burning and Transplanting Indian Corn. The summer, and the fall also, have been *remarkably dry* in *Long Island*, much more dry than is usual.

The grass has been very short indeed. A sort of Grass-hopper, or cricket, has eaten up a considerable part of the grass and of all vegetables, the leaves of which have come since the month of June. I am glad, that this has been the case; for I now know what a farmer may do in the *worst of years;* and, when I consider what the summer has been, I look at my Cabbages and Ruta Baga with surprize as well as with satisfaction.

Cabbages.

166. I had some hogs to keep, and, as my Swedish Turnips (Ruta Baga) would be gone by July, or before, I wished them to be succeeded by cabbages. I made a *hot-bed* on the *20th of March*, which ought to have been made more than a month earlier; but, I had been in Pennsylvania, and did not return home till the *13th of March.* It requires a little time to mix and turn the dung in order to prepare it for a hot-bed; so that mine was not a very good one; and then my *frame* was hastily patched up, and its covering consisted of some old *broken* sashes of windows. A very shabby concern; but, in this bed I sowed *cabbages* and *cauliflowers.* The seed came up, and the plants, though standing too thick,

grew pretty well. From this bed, they would, if I had had time, been transplanted into another, at about two and a half or three inches apart. But, such as they were, very much drawn up, I began planting them out as soon as they were about four inches high.

167. It was the *12th of May* before they attained this height, and I then began planting them out in a piece of ground, pretty good, and deeply ploughed by oxen. My cauliflowers, of which there were about three thousand, were *too late* to *flower*, which they never will do, unless the flower have begun to shew itself before the great heat comes. However, these plants grew *very large*, and afforded a great quantity of food for pigs. The outside leaves and stems were eaten by sows, store-pigs, a cow, and some oxen; the hearts, which were very tender and nearly of the Cauliflower-taste, were boiled in a large cast-iron caldron, and, mixed with a little rye-meal, given to sows and young pigs. I should suppose, that these three thousand plants weighed twelve hundred pounds, and they stood upon about half an acre of land. I gave these to the animals *early in July*.

168. The *Cabbages*, sown in the bed, consisted partly of Early Yorks, the seed of which had been sent me along with the Cauli-

flower seed, from England, and had reached me at Harrisburgh in Pennsylvania; and partly of plants, the seed of which had been given me by Mr. JAMES PAUL, Senior, of Bustleton, as I was on my return home. And this gave me a pretty good opportunity of ascertaining the fact as to the *degenerating of cabbage seed.* Mr. Paul, who attended very minutely to all such matters; who took great delight in his garden; who was a *reading* as well as a practical farmer, told me, when he gave me the seed, that it would not produce loaved cabbages so early as my own seed would; for, that, though he had always selected the earliest heads for seed, the seed degenerated, and the cabbages regularly came to perfection later and later. He said, that he never should save cabbage seed himself; but, that it was such *chance-work* to buy of seedsmen, that he thought it best to save some at any rate. In this case, *all* the plants from the English seed produced solid loaves by the 24th of June, while, from the plants of the Pennsylvania seed, we had not a single solid loaf till the 28th of July, and, from the chief part of them, not till mid-August.

169. This is a great matter. Not only have you the food earlier, and so much earlier, from the genuine seed, but your ground is occupied

so much less time by the plants. The plants very soon shewed, by their appearance, what would be the result; for, on the 2nd of June, Miss Sarah Paul, a daughter of Mr. James Paul, saw the plants, and while those from the English seed were even then beginning to loave, those from her father's seed were nothing more than bunches of wide spreading leaves, having no appearance of forming a head. However, they succeeded the plants from the English seed; and, the whole, besides what were used in the House, were given to the animals. As many of the *white* loaves as were wanted for the purpose were boiled for sows and small pigs, and the rest were given to lean pigs and the horn-cattle: and a fine resource they were; for, so dry was the weather, and the devastations of the grass-hoppers so great, that we had scarcely any grass in any part of the land; and, if I had not had these cabbages, I must have resorted to Indian Corn, or Grain of some sort.

170. But, these spring-cabbage plants were to be *succeeded* by others, to be eaten in September and onwards to January. Therefore, on the 27th of May, I sowed in the natural ground eleven sorts of cabbages, some of the seed from England and some got from my friend, Mr. PAUL. I have noticed the *extreme*

drought of the season. Nevertheless, I have now about two acres of cabbages of the following description. Half an acre of the *Early Salisbury* (earliest of all cabbages) and *Early York;* about 3 quarters of an acre of the *Drum-head* and other late cabbages; and about the same quantity of *Green Savoys.* The first class are fully loaved, and bursting: with these I now feed my animals. These will be finished by the time that I cut off my Swedish Turnip Greens, as mentioned in Part I. Paragraph 136. Then, about mid-December, I shall feed with the second class, the Drum-heads and other late Cabbages. Then, those which are not used before the *hard frosts* set in, I shall *put up* for use through the month of January.

171. Aye! *Put them up;* but how? No scheme that industry or necessity ever sought after, or that experience ever suggested, with regard to the preserving of cabbages, did I leave untried last year; and, in every scheme but one I found some inconvenience. Taking them up and replanting them closely in a sloping manner and covering them with straw; putting them in pits; hanging them up in a barn; turning their heads downwards and covering them with earth, leaving the roots sticking up in the air: in short every scheme, except one, was attended with great labour,

and some of them forbade the hope of being able to preserve any considerable quantity; and this one was as follows: I made a sort of *land* with the plough, and made it pretty level at top. Upon this land I laid some straw. I then took the cabbages, turned them upside down, and placed them (first taking off all decayed leaves) about six abreast upon the straw. Then covered them, not very thickly, with leaves raked up in the woods, flinging now and then a little dirt (boughs of any sort would be better) to prevent the leaves from being carried off by the wind. So that, when the work was done, the thing was a bed of leaves with cabbage-roots sticking up through it. I only put on enough leaves to *hide all the green.* If the frost came and prevented the taking up of the cabbages, roots and all, they might be cut off *close to the ground.* The root, I dare say, is of no use in the preservation. In the months of *April* and *May*, I took cabbages of all sorts from this *land* perfectly good and fresh. The quantity, preserved thus, was small. It might amount to 200 cabbages. But, it was quite sufficient for the purpose. Not only did the cabbages *keep* better in this, than in any other way, but there they were, *at all times, ready.* The frost had *locked up* all those which were covered with earth, and those which lay with heads upwards and their roots in the ground *were rotting*. But, to this *land* I could have gone at any time, and have brought away, if the quantity had been large, a waggon load in ten minutes. If they had been *covered with snow* (no matter how deep) by uncovering twenty feet in length (a work of little labour) half a ton of cabbages would have been got at. This year, thinking that my *Savoys*, which are, at once, the best in quality and best to keep, of all winter cabbages, may be of use to send to New York, I have planted them between rows of *Broom-Corn*. The Broom-Corn is in *rows*, eight feet apart. This enabled us to plough deep between the Broom-Corn, which, though in poor land, has been very fine. The heads are cut off; and now the *stalks* remain to be used as follows: I shall make *lands* up the piece, cut off the stalks and lay them, first a layer longways and then a layer crossways, upon the *lands*. Upon these I shall put my Savoys turned upside down; and, as the stalks will be more than sufficient for this purpose, I shall lay some of them *over*, instead of dirt or boughs, as mentioned before. Perhaps the *leaves* of the Broom-Corn, which are lying about in great quantities, may suffice for covering. And, thus, all the materials for the work are upon the spot.

172. In quitting this matter, I may observe, that, to cover cabbages thus, in gardens as well as fields, would, in many cases, be of great use in *England*, and of still more use in Scotland. Sometimes, a quick succession of frost, snow and thaw will completely *rot* every *loaved* cabbage even in the South of England. Indeed no reliance is placed upon cabbages for use, as cattle-food, later than the month of *December*. The bulk is so large that a protection by *houses* of any sort cannot be thought of. Besides, the cabbages, put together in large masses would *heat* and quickly rot. In *gentlemen*'s gardens, indeed, cabbages are put into houses, where they are hung up by the heads. But, they *wither* in this state, or they soon *putrefy* even here. By adopting the mode of preserving, which I have described above, all these inconveniences would be avoided. Any quantity might be preserved either in fields or in gardens at a very trifling expence, compared with the bulk of the crop.

173. As to the application of my Savoys, and part of the Drum-heads, too, indeed, if I find cabbages very dear, at New York, in winter, I shall send them; if not, there they are for my cattle and pigs. The weight of them will not be less, I should think, than *ten tons*. The plants were put out *by two men in*

one day; and I shall think it very hard if two men do not put the whole completely up *in a week.* The Savoys are very fine. A little too late planted out; but still very fine; and they were planted out under a burning sun and without a drop of rain for weeks afterwards. So far from taking any *particular* pains about these Savoys, I did not see them planted, and I never saw them for *more than two months* after they were planted. The ground for them was prepared thus: the ground, in each interval between the Broom-Corn, had been, some little time before, ploughed *to* the rows. This left a *deep furrow* in the middle of the interval. Into this furrow I put the manure. It was a mixture of good mould and dung from pig-styes. The waggon went up the interval, and the manure was drawn out and tumbled into the furrow. Then the plough went twice on each side of the furrow, and turned the earth over the manure. This made a *ridge,* and upon this ridge the plants were planted as quickly after the plough as possible.

174. Now, then, what is the *trouble;* what is the *expence,* of all this? The *seed* was excellent. I do not recollect ever having seen so large a piece of the cabbage kind with so few spurious plants. But, though *good* cabbage seed is of *high price,* I should suppose, that

the seed did not cost me *a quarter of a dollar.* Suppose, however, it had cost *ten quarters of a dollar;* what would that have been, compared to the worth of the crop? For, what is the worth of *ten tons* of green, or moist food, in the month of March or April?

175. The Swedish Turnip is, indeed, still *more conveniently* preserved, and is a *richer* food; but, there are some reasons for making *part* of the year's provision to consist of cabbages. As far as a thing may depend on *chance*, two chances are better than one. In the *summer and fall*, cabbages get *ripe*, and, as I have observed, in Part I. Paragraph 143, the Ruta Baga (which we will call *Swedish Turnip* for the future) is not *so good* 'till it be *ripe;* and is a great deal better when kept 'till February, than when used in December. This matter of *ripeness* is worthy of attention. Let any one eat a piece of *white cabbage;* and then eat a piece of the same sort of cabbage *young and green.* The first he will find *sweet*, the latter *bitter.* It is the same with *Turnips*, and with all roots. There are some apples, wholly uneatable 'till *kept a while*, and then delicious. This is the case with the Swedish Turnip. Hogs will, indeed, always *eat it*, young or old; but, it is not nearly so good early, as it is when kept 'till February. However, in default of other things, I would feed with it even in November.

176. For these reasons I would have my due proportion of cabbages, and I would always, if possible, have some Green Savoys; for, it is, with cabbages, too, not only *quantity* which we ought to think of. The Drum-head, and some others, are called *cattle-cabbage;* and hence, in England, there is an idea, that the more delicate kinds of cabbage are *not so good* for cattle. But, the fact is, that they are as much *better* for cattle, than the coarse cabbages are, as they are better for us. It would be strange indeed, that, reversing the principle of our general conduct, we should give cabbage of the best quality to cattle, and keep that of the worst quality for ourselves. In London, where taxation has kept the streets as clear of bits of meat left on bones as the hogs endeavour to keep the streets of New York, there are people who go about selling "*dog's meat.*" This consists of boiled garbage. But, it is not pretended, I suppose, that dogs will not eat roast-beef; nor, is it, I suppose, imagined, that they would not *prefer* the roast-beef, if they had their choice? Some people pretend, that garbage and carrion are *better* for dogs than beef and mutton are. That is to say, it is *better for us,* that they should live upon things, which we ourselves loath, than that they should share with us. Self-interest is, but too frequently, a miserable logician.

177. However, with regard to cattle, sheep, and pigs, as we intend to eat *them*, their claim to our kindness is generally more particularly and impartially listened to than that of the poor dogs; though that of the latter, founded, as it is, on their sagacity, their fidelity, their real utility, as the guardians of our folds, our home-steads and our houses, and as the companions, or, rather, the givers, of our healthful sports, is ten thousand times more strong, than that of animals which live to eat, sleep, and grow fat. But, to return to the cabbages, the fact is, that all sorts of animals, which will eat them at all, like the most delicate kinds best; and, as some of these are also the *earliest* kinds, they ought to be cultivated for cattle. Some of the larger kinds may be cultivated *too;* but, they cannot be got ripe till the fall of the year. Nor is the difference in the *weight* of the crop so great as may be imagined. On the same land, that will bear a Drum-head of *twenty pounds*, an Early York, or Early Battersea will weigh *four pounds;* and these may be *fifteen inches* asunder in the row, while the Drum-head requires *four feet*. Mind, I always suppose the *rows* to be *four feet* apart, as stated in the First Part of this work, and for the reasons there stated. Besides the advantages of having some cabbages *early*, the early ones remain so little a

time upon the ground. Transplanted Swedish Turnips, or Buckwheat, or late Cabbages, especially Savoys, may always follow them the same year upon the same land. My early cabbages, this year, have been followed by a second crop of the same, and now (mid-November) they are hard and white and we are giving them to the animals.

178. There is a convenience attending cabbages, which attends no other of the cattle-plants, namely, that of raising the *plants* with very little trouble and upon a small bit of ground. A *little bed* will give plants for an acre or two. The expence of *seed*, even of the dearest kinds, is a mere trifle, not worth any man's notice.

179. For these reasons I adhere to cabbages as the companion crop of Swedish Turnips. The Mangel Wurzel *is long in the ground.* In seasons of great drought, it comes up *unevenly.* The weeds get the start of it. Its tillage must begin before it hardly shews itself. It is of the nature of the Beet, and it requires the care which the Beet requires. The same may be said of Carrots and Parsnips. The cabbage, until it be fit to plant out, occupies hardly any ground. An hours work cleans the bed of weeds; and there the plants are always ready, when the land is made ready. The Mangel

Wurzel *root*, if quite *ripe*, is richer than a white loaved cabbage; but, it is not more easily preserved, and will not produce a larger crop. Cattle will eat the *leaves*, but hogs will not, when they can get the *leaves* of cabbages. Nevertheless, *some* of this root may be cultivated. It will *fat an ox* well; and it will *fat sheep* well. Hogs will do well on it in winter. I would, if I were a settled farmer, have *some* of it; but, it is not a thing upon which I would place my *dependence*.

180. As to the time of sowing cabbages, the first sowing should be in a hot-bed, so as to have the plants *a month old when the frost leaves the ground.* The second sowing should be *when the natural ground has become warm enough to make the weeds begin to come up freely.* But, seed-beds of cabbages, and, indeed, of every thing, should be *in the open:* not under *a fence*, whatever may be the aspect. The plants are sure to be weak, if sown in such situations. They should have the air coming freely to them in every direction. In a hot-bed, the seed should be sown in rows, three inches apart, and the plants might be thinned out to one in a quarter of an inch. This would give about *ten thousand plants* in a bed *ten feet long, and five wide.* They will stand thus to get to a tolerable size without injuring each

other, if the bed be well managed as to *heat* and *air*. In the open ground, where room is plenty, the rows may be a foot apart, and the plants two inches apart in the rows. This will allow of *hoeing*, and here the plants will grow very finely. Mind, a *large* cabbage plant, as well as a large turnip plant, is *better* than a small one. All will grow, if well planted; but the large plant will grow best, and will, in the end, be the finest cabbage.

181. We have a way, in England, of greatly improving the plants; but, I am almost afraid to mention it, lest the American reader should be *frightened* at the bare thought of the *trouble*. When the plants, in the seed-bed, have got leaves about an inch broad, we take them up, and transplant them in *fresh ground*, at about *four inches apart each way*. Here they get *stout* and *straight*; and, in about three weeks time, we transplant them again into the ground where they are to come to perfection. This is called *pricking out*. When the plant is removed the second time, it is found to be furnished with new roots, which have shot out of the butts of the long tap, or forked roots, which proceeded from the seed. It, therefore, *takes again* more readily to the ground, and has some earth adhere to it in its passage. One hundred of pricked-out plants are always look-

ed upon as worth three hundred from the seed-bed. In short, no man, in England, unless he be extremely negligent, ever *plants out* from the seed-bed. Let any farmer try this method with only a score of plants. He may do it with *three minutes*' labour. Surely, he may spare *three minutes*, and I will engage, that, if he treat these plants afterwards as he does the rest, and, if all be treated well, and the crop a *fair* one, the three minutes will give him fifty pounds weight of any of the larger sorts of cabbages. Plants are *thus* raised, then taken up and tied neatly in bundles, and then brought out of Dorsetshire and Wiltshire, and sold in Hampshire for *three-pence* (about *six cents*) a hundred. So that it cannot require the heart of a lion to encounter the *labour* attending the raising of a few thousands of plants.

182. However, my plants, this year, have all gone into the field from the seed-bed; and, in so fine a climate, it may do very well; only great care is necessary to be taken to see that they be not *too thick* in the seed-bed.

183. As to the preparation of the land, as to the manuring, as to the *distance* of the rows from each other, as to the act of planting, and as to the after culture, all are the same as in the case of transplánted Swedish Turnips; and, therefore, as to these matters, the reader has

seen enough in Part I. There is one observation to make, as to the *depth* to which the plant should be put into the ground. It should be placed so deep, that the stems of the outside leaves be *just clear of the ground;* for, if you put the plant deeper, the rain will wash the loose earth in amongst the stems of the leaves, which will make an open poor cabbage; and, if the plant be placed so low as for the *heart to be covered with dirt*, the plant, though it will *live*, will come to nothing. Great care must, therefore, be taken as to this matter. If the stems of the plants be *long*, roots will burst out nearly all the way up to the surface of the earth.

184. The distances at which cabbages ought to stand *in the rows* must depend on the sorts. The following is nearly about the mark. Early Salisbury *a foot;* Early York *fifteen inches;* Early Battersea *twenty inches;* Sugar Loaf two feet; Savoys two feet and a half; and the Drum-head, Thousand-headed, Large Hollow, Ox cabbage, all *four feet*.

185. With regard to the *time of sowing* some more ought to be said; for, we are not here, as in England, confined within four or five degrees of latitude. Here some of us are living in fine, warm weather, while others of us are living amidst snows. It will be better, there-

fore, in giving opinions about *times*, to speak of *seasons*, and not of months and days. The country people, in England, go, to this day, many of them, at least, by the *tides;* and, what is supremely ridiculous, they go, in some cases, by the *moveable tides*. My gardener, at Botley, very reluctantly obeyed me, one year, in sowing green Kale when I ordered him to do it, because *Whitsuntide* was not come, and that, he said, was the *proper season*. "But," said I, "Robinson, Whitsuntide comes *later* this year "than it did last year." "*Later*, Sir," said he, "how can that be?" "Because," said I, "it "depends upon *the moon* when Whitsuntide "shall come." "The *moon!*" said he: "what "*sense* can there be in that?" "Nay," said I, "I am sure I cannot tell. That is a matter "far beyond my learning. Go and ask Mr. "BAKER, the Parson, He ought to be able to "tell us; for he has a tenth part of our gar-"den stuff and fruit." The Quakers here cast all this rubbish away; and, one wonders how it can possibly be still cherished by any portion of an enlightened people. But, the truth is, that men do not *think for themselves* about these matters. Each succeeding generation tread in the steps of their fathers, whom they loved, honoured and obeyed. They take all upon trust. Gladly save themselves the trouble of thinking about things of not immediate interest. A desire to avoid the reproach of being irreligious induces them to practise an outward conformity. And thus have priestcraft with all its frauds, extortions, and immoralities, lived and flourished in defiance of reason and of nature.

186. However, as there are no farmers in America quite foolish enough to be ruled by the *tides* in sowing and reaping, I hurry back from this digression to say, that I cannot be expected to speak of *precise times* for doing any work, except as relates to the latitude in which I live, and in which my experiments have been made. I have cultivated a garden at Frederickton in the *Province of New Brunswick*, which is in latitude about *forty-eight;* and at Wilmington in Delaware State, which is in latitude about *thirty-nine*. In *both these places* I had as fine cabbages, turnips, and garden things of all the hardy sorts, as any man need wish to see. Indian Corn grew and ripened well in fields at Frederickton. And, of course, the summer was sufficient for the perfecting of all plants for cattle-food. And, how *necessary* is this food in Northern Climates! More to the Southward than Delaware State I have not been; but, in those countries the farmers have to pick and choose. They have two Long

Island summers and falls, and three English, in *every year.*

187. According to these various circumstances men must form their judgment; but, it may be of some use to state the *length of time*, which is required to bring each sort of cabbage to perfection. The following sorts are, it appears to me, all that can, in any case, be necessary. I have put against each nearly the time, that it will require to bring it to *perfection*, from the time of *planting out* in the places where the plants are to stand to come to perfection. The plants are supposed to be of a good size when *put out*, to have stood sufficiently *thin* in the seed-bed, and to have been kept clear from weeds in that bed. They are also supposed to go into ground well prepared.

Sort	Time
Early Salisbury	Six weeks.
Early York	Eight weeks.
Early Battersea	Ten weeks.
Sugar Loaf	Eleven weeks.
Late Battersea	Sixteen weeks.
Red Kentish	Sixteen weeks.
Drum-head	Five months.
Thousand-headed	Five months.
Large hollow	Five months.
Ox cabbage	Five months.
Savoy	Five months.

188. It should be observed, that Savoys, which are so very rich in winter, are not so good, till they have been *pinched by frost*. I have put *red* cabbage down as a sort to be cultivated, because they are as good as the white of the same size, and because it may be convenient, in the farmer's family, to have some of them. The *thousand-headed* is of prodigious produce. You pull off the heads, of which it bears a great number at first, and others come; and so on for months, if the weather permit; so that this sort does not take five months to bring its *first* heads to perfection. When I say *perfection*, I mean quite *hard;* quite *ripe*. However, this is a *coarse* cabbage, and requires great room. The *Ox-cabbage* is coarser than the *Drum-head*. The *Large hollow* is a very fine cabbage; but it requires very good land. Some of all the sorts would be best; but, I hope, I have now given information enough to enable any one to form a judgment correct enough to *begin* with. Experience will be the best guide for the future. An *ounce* of each sort of seed would, perhaps, be enough; and the cost is, when compared with the object, too trifling to be thought of.

189. Notwithstanding all that I have said, or can say, upon the subject of cabbages, I am very well aware, that the extension of the cul-

tivation of them, in America, will be a work of *time.* A proposition to do any thing *new,* in so common a calling as agriculture, is looked at with suspicion; and, by some, with feelings not of the kindest description; because it seems to imply an imputation of *ignorance* in those to whom the proposition is made. A little reflection will, however, suppress this feeling in men of sense; and, those who still entertain it may console themselves with the assurance, that no one will desire to *compel* them to have stores of green, or moist, cattle-food in winter. To be *ashamed* to be taught is one of the greatest of human follies; but, I must say, that it is a folly less prevalent in America than in any other country with which I am acquainted.

190. Besides the disposition to reject novelties, this proposition of mine has *books* to contend against. I read, last fall, in an American Edition of the Encyclopædia Britannica, "*greatly enlarged and improved,*" some observations on the culture of cabbages as cattle-food, which were well calculated to deter a reader of that book from attempting the culture. I do not recollect the *words;* but, the substance was, that this plant *could not be cultivated to advantage by the farmer* IN AMERICA. This was the more provoking to me, as I had,

at that moment, so fine a piece of cabbages in *Long Island.* If the American Editor of this work had given his readers the bare, *unimproved,* Scotch Edition, the reader would have there seen, that, in England and Scotland, they raise *sixty-eight tons* of cabbages (*tons* mind) upon *an acre;* and that the whole expence of an acre, exclusive of rent, is *one pound, fourteen shillings and a penny;* or *seven dollars and seventy-five cents.* Say that the expence in America is double and the crop one half, or *one fourth,* if you like. Where are *seventeen tons* of green food in winter, or even in summer, to be got for *sixteen dollars;* Nay, where is that quantity, of such a quality, to be got for *fifty dollars?* The Scotch Edition gives an account of *fifty-four tons* raised on an acre where the land was worth only *twelve shillings* (less than *three dollars*) an acre. In *fairness,* then, the American Editor should have given to his agricultural readers what the Scotchman had said upon the subject. And, if he still thought it right to advise the American farmers not to think of cabbages, he should, I think, have offered them some, at least, of the *reasons* for his believing, that that which was obtained in such abundance in England and Scotland, was not to be obtained to any profit at a l here. What! will not this immense region furnish a climate, for this purpose, equal even to Scotland, where an *oat* will hardly ripen; and where the crop of that miserable grain is sometimes harvested amidst ice and snow! The proposition is, upon the face of it, an absurdity; and my experience proves it to be false.

191. This book says, if I recollect rightly, that the culture has been *tried*, and has *failed.* Tried? How tried? That cabbages, and most beautiful cabbages *will grow*, in all parts of America, every farmer knows; for he *has* them in his garden, or sees them, every year, in the gardens of others. And, if they will grow in *gardens*, why not in *fields?* Is there common sense in supposing, that they will not grow in a piece of land, because it is not *called a garden?* The Encyclopædia Britannica gives an account of *twelve acres* of cabbages, which would keep "*forty-five oxen* and *sixty sheep* " for *three months;* improving them as much as " the grass in the best months in the year (in " England) May, June, and July." Of these large cabbages, being at four feet apart in the rows, one man will easily plant out *an acre in a day.* As to the *seed-bed*, the labour of that is nothing, as we have seen. Why, then, are men frightened at the *labour?* All but the mere act of planting is performed by oxen or

horses; and they never complain of "*the labour.*" The labour of an acre of cabbages is *not half so much as that of an acre of Indian Corn.* The bringing in of the crop and applying it are not more expensive than those of the corn. And will any man pretend, that an acre of good cabbages is not worth three times as much as a crop of good corn? Besides, if *early* cabbages, they are off and leave the land for transplanted Swedish Turnips, for Late Cabbages, or for Buckwheat; and, if *late* cabbages, they come after early ones, after wheat, rye, oats, or barley. This is what takes place even in England, where the fall is so much shorter, as to growing weather, than it is in Long Island, and, of course, all the way to Georgia. More to the North, in the latitude of Boston, for instance, two crops of early cabbages will come upon the same ground; or a crop of early cabbages will follow any sort of grain, except Buckwheat.

192. In concluding this Chapter I cannot help strongly recommending farmers who may be disposed to try this culture, to try it *fairly.* That is to say, to employ *true seed, good land,* and *due care;* for, as "men do not gather "grapes from thorns, nor figs from thistles," so they do not harvest cabbages from stems of rape. Then, as to the land, it must be made

good and rich, if it be not in that state already; for a cabbage will not be fine, where a white Turnip will; but as the quantity of land, wanted for this purpose, is comparatively very small, the land may easily be made rich. The after-culture of cabbages is trifling. No weeds to plague us with *hand*-work. Two good ploughings, at most, will suffice. But ploughing after planting out is necessary; and, besides, it leaves the ground in so fine a state. The trial may be on a small scale, if the farmer please. Perhaps it were best to be such. But, on whatever scale, let the *trial* be a *fair trial.*

193. I shall speak again of the *use* of cabbages, when I come to speak of *Hogs* and *Cows.*

CHAP. IV.

EARTH-BURNING, 1818.

194. In paragraphs 99, 100, and 101, I spoke of a mode of procuring *manure* by the burning of *earth*, and I proposed to try it this present year. This I have now done, and I proceed to give an account of the result.

195. I have tried the efficacy of this manure on Cabbages, Swedish Turnips, Indian Corn, and Buckwheat. In the three former cases the Ashes were put into the furrow and the earth was turned over them, in the same way that I have described, in Paragraph 177, with regard to the manure for Savoys. I put at the rate of about *twenty tons* weight to an acre. In the case of the Buckwheat, the Ashes were spread out of the waggon upon a little strip of land on the out-side of the piece. They were *thickly* spread; and it might be, that the proportion exceeded even *thirty tons* to the acre. But, upon the part where the ashes were spread, the Buckwheat was three or four times as good as upon the land adjoining. The land was *very poor*. It bore Buckwheat *last* year, without any manure.

It had two good ploughings then, and it had two good ploughings again this year, but had no manure, except the part above-mentioned and one other part at a great distance from it. So that the trial was very fair indeed.

196. In every instance the ashes produced *great effect;* and I am now quite certain, that *any crop* may be raised with the help of this manure; that is to say, any *sort* of crop; for, of dung, wood-ashes, and earth-ashes, when all are ready upon the spot, without *purchase* or *carting from a distance,* the two former are certainly to be employed in preference to the latter, because a smaller quantity of them will produce the same effect, and, of course, the application of them is less expensive. But, in taking to a farm unprovided with the two former; or under circumstances which make it profitable to *add* to the land under cultivation, what can be so convenient, what so cheap, as ashes procured in this way?

197. A near neighbour of mine, Mr. DAYREA, sowed a piece of Swedish Turnips, broad-cast, in June, this year. The piece was near a wood, and there was a great quantity of *clods* of a *grassy* description. These he burnt into *ashes,* which ashes he spread over one half of the piece, while he put *soaper's ashes* over the other part of the piece. I saw the turnips in

October; and there was no visible difference in the two parts, whether as to the vigourousness of the plants or the bulk of the turnips. They were sown broad-cast, and stood unevenly upon the ground. They were harvested a month ago (it is now 26 November), which was *a month too early*. They would have been a third, at least, more in bulk, and much better in quality, if they had remained in the ground until now. The piece was 70 paces long and 7 paces wide; and, the reader will find, that, as the piece produced *forty bushels*, this was at the rate of *four hundred bushels to the acre*.

198. What quantity of *earth ashes* were spread on this piece it is impossible to ascertain with precision; but, I shall suppose the quantity to have been very large indeed in proportion to the surface of the land. Let it be four times the quantity of the soaper's ashes. Still, the one was made upon the spot, at, perhaps, a tenth part of the cost of the other; and, as such ashes can be made upon any farm, there can be no reason for not *trying* the thing, at any rate, and which *trying* may be effected upon so small a scale as not to exceed in expence a half of a dollar. I presume, that many farmers will try this method of obtaining manure; and, therefore, I will describe how the burning is effected.

199. There are two ways of producing ashes from earth: the one in heaps upon the ground, and the other within walls of turf, or earth. The first, indeed, is the burning of *turf*, or *peat*. But, let us see how it is done.

200. The surface of the land is taken off to a depth of two or three inches, and turned the earth side uppermost to *dry*. The land, of course, is covered with grass, or heath, or something the roots of which hold it together, and which makes the part taken off take the name of *turf*. In England, this operation is performed with a *turf-cutter*, and by hand. The turfs are then taken, or a part of them, at least, and placed on their *edges*, leaning against each other, like the two sides of the roof of a house. In this state they remain, 'till they are dry enough to burn. Then the burning is begun in this way. A little straw and some dry sticks, or any thing that will make a trifling fire, is lighted. Some little bits of the turf are put to this. When the turf is on fire, more bits are carefully put round against the openings whence the smoke issues. In the course of a day or two the heap grows large. The burning keeps working on the inside, though there never appears any *blaze*. Thus the field is *studded* with heaps. After the *first* fire is got to be of considerable bulk, no straw is wanted

for other heaps, because a good shovel full of *fire* can be carried to light other heaps; and so, until all the heaps are lighted. Then the workman goes from heap to heap, and carries the turf to all, by degrees, putting some to each heap every day or two, until all the field be burnt. He takes care to keep in the smoke as much as possible. When all the turf is put on, the field is left; and, in a week or two, whether it rain or not, the heaps are *ashes* instead of earth. The ashes are afterwards spread upon the ground; the ground is ploughed and sowed; and this is regarded as the very best preparation for a crop of turnips.

201. This is called "*paring and burning*." It was introduced into England by the *Romans*, and it is strongly recommended in the First Georgic of Virgil, in, as Mr. TULL shows, very fine poetry, very bad philosophy, and still worse logic. It gives three or four crops upon even poor land; but, it *ruins* the land for an age. Hence it is, that *tenants*, in England, are, in many cases, *restrained* from paring and burning, especially towards the close of their leases. It is the Roman husbandry, which has always been followed, until within a century, by the French and English. It is implicitly followed in France to this day; as it is by the great mass of common farmers in

England. All the foolish country sayings about *Friday* being an *unlucky day* to begin any thing fresh upon; about the *noise of Geese* foreboding bad weather; about the *signs of the stars;* about the influence of the *moon* on animals: these, and scores of others, equally ridiculous and equally injurious to true philosophy and religion, came from the Romans, and are inculcated in those books, which pedants call "*classical*," and which are taught to "young *gentlemen*" at the universities and in academies. Hence, too, the foolish notions of sailors about *Friday*, which notions very often retard the operations of commerce. I have known many a farmer, when his wheat was dead ripe, put off the beginning of harvest from Thursday to Saturday, in order to avoid *Friday*. The *stars* save hundreds of thousands of lambs and pigs from sexual degradation at so early an age as the operation would otherwise be performed upon them. These heathen notions still prevail even in America as far as relates to this matter. A neighbour of mine in Long Island, who was to operate on some pigs and lambs for me, begged me to put the thing off for a while; for, that the *Almanac* told him, that the *signs* were, just then, as *unfavourable* as possible. I begged him to proceed, for that I set all *stars* at defiance. He very kindly complied, and had the pleasure to see, that every pig and lamb did well. He was surprized when I told him, that this mysterious matter was not only a bit of *priest-craft*, but of *heathen* priest-craft, cherished by priests of a more modern date, because it tended to bewilder the senses and to keep the human mind in subjection. "What a thing it is, Mr. "Wiggins," said I, "that a cheat practised upon "the pagans of Italy, two or three thousand "years ago, should, by almanac-makers, be "practised on a sensible farmer in America!" If priests, instead of preaching so much about mysteries, were to explain to their hearers the origin of cheats like this, one might be ready to allow, that the wages paid to them were not wholly thrown away.

202. I make no apology for this digression; for, if it have a tendency to set the minds of only a few persons on the track of detecting the cheatery of priests, the room which it occupies will have been well bestowed.

203. To return to *paring and burning*; the reader will see with what ease it might be done in America, where the *sun* would do more than half the work. Besides the *paring* might be done with the *plough*. A sharp shear, going shallow, could do the thing perfectly well. Cutting *across* would make the *sward* into turfs.

204. So much for *paring and burning*. But, what I recommend is, not to burn the land which is to be cultivated, but *other earth*, for the purpose of getting ashes to be brought on the land. And this operation, I perform thus: I make a circle, or an oblong square. I cut *sods* and build a wall all round, three feet thick and four feet high. I then light a fire in the middle with straw, dry sticks, boughs, or such like matter. I go on making this fire larger and larger till it extends over the whole of the bottom of the pit, or kiln. I put on roots of trees or any rubbish wood, till there be a good thickness of strong coals. I then put on the *driest* of the clods that I have ploughed up round about so as to cover all the fire over. The earth thus put in will burn. You will see the smoke coming out at little places here and there. Put more clods wherever the smoke appears. Keep on thus for a day or two. By this time a great mass of fire will be in the inside. And now you may dig out the clay, or earth, any where round the kiln, and fling it on without ceremony, always taking care *to keep in the smoke;* for, if you suffer that to continue coming out at any one place, a hole will soon be made; the main force of the fire will draw to that hole; a blaze, like that of a volcano will come out, and the fire will be extinguished.

205. A very good way, is, to put your finger into the top of the heap here and there; and if you find the fire *very near*, throw on more earth. Not *too much at a time;* for that weighs too heavily on the fire, and keeps it back; and, at *first*, will put it partially out. You keep on thus augmenting the kiln, till you get to the top of the walls, and then you may, if you like, raise the walls, and still go on. No rain will affect the fire when once it is become strong.

206. The principle is to *keep out air*, whether at the top or the sides, and this you are sure to do, if you *keep in the smoke.* I burnt, this last summer, about thirty waggon loads in one round kiln, and never saw the smoke at all after the first four days. I put in my finger to try whether the fire was near the top; and when I found it approaching, I put on more earth. Never was a kiln more completely burnt.

207. Now, this may be done on the skirt of any wood, where the matters are all at hand. This mode is far preferable to the *above-ground* burning in *heaps.* Because, in the first place, there the materials must be *turf*, and dry turf; and, in the next place, the *smoke escapes there*, which is the finest part of burnt matter. *Soot*, we know well, is more powerful than ashes;

and, soot is composed of the *grossest part of the smoke.* That which flies out of the chimney is the best part of all.

208. In case of a want of *wood* wherewith to begin the fire, the fire may be lighted precisely as in the case of *paring and burning.* If the kiln be large, the oblong square is the best figure. About *ten feet wide,* because then a man can fling the earth easily over every part. The mode they pursue in England, where there is no *wood,* is to make a sort of building in the kiln with turfs, and leave air-holes at the corners of the walls, till the fire be well begun. But this is tedious work; and, in this country wholly unnecessary. Care must, however, be taken, that the fire be well lighted. The matter put in *at first* should be such as is of the lightest description; so that a *body of earth on fire* may be obtained, before it be too heavily loaded.

209. The burning being completed, having got the quantity you want, let the kiln remain. The fire will continue to work, 'till all is ashes. If you want to *use* the ashes sooner, open the kiln. They will be cold enough to remove in a week.

210. Some persons have *peat,* or bog earth. This may be burnt like common earth, in kilns, or *dry,* as in the paring and burning method.

Only, the *peat* should be cut out in the *shape of bricks*, as much longer and bigger as you find convenient, and set up to dry, in the same way that bricks are set up to dry previous to the burning. This is the *only fuel* for houses in some parts of England. I myself was nursed and brought up without ever seeing any other sort of fire. The ashes used, in those times, to be sold for *four pence sterling a bushel*, and were frequently carried, after the purchase, to a distance of ten miles, or more: At this time, in my own neighbourhood, in Hampshire, peat is burnt in large quantities for the ashes, which are sold, I believe, as high as *sixpence* sterling a bushel, and carried to a distance even of twenty miles in some cases.

211. Nevertheless it is certain, that these ashes are not equally potent upon every sort of soil. We do not use them much at Botley, though upon the spot. They are carried away to the higher and poorer lands, where they are *sown by hand* upon *clover* and *sain-foin*. An excellent farmer, in this Island, assures me, that he has tried them in various ways, and never found them to have effect. So say the farmers near Botley. But, there is no harm in making a *trial*. It is done with a mere nothing of expence. A yard square in a garden is quite sufficient for the experiment.

212. With respect to earth-ashes, burnt in kilns, *keeping in the smoke*, I have proved their great good effect; but, still, I would recom mend *trying* them upon a small scale. However, let it be borne in mind, that the proportion to the acre ought to *be large*. Thirty good tons to an acre; and why may it not be such, seeing that the expence is so trifling?

CHAP. V.

TRANSPLANTING INDIAN CORN.

213. I WAS always of opinion, that this would be the best mode, under certain circumstances, of dealing with this crop. The *spring*, in this part of America, and further to the North, is but *short*. It is nearly winter 'till it is summer. The labours of the year are, at this season, very much *crowded*. To plant the grains of the Indian Corn over a whole field requires previous ploughing, harrowing, marking, and manuring. The consequence is, that, as there are so many other things to do, something is but too often badly done.

214. Now, if this work of Corn planting could be postponed to the 25th of June (for this Island) instead of being performed on, or about the 15th of May, how well the ground might be prepared by the 25th of June! This can be done only by transplanting the plants of the Corn. I was resolved to try this; and so confident was I that it would succeed, that I had made some part of my preparations for *six acres*.

215. I sowed the seed at about three inches apart, in beds, on *the 20th of May.* The plants stood in the beds (about 15 perches of ground) till the *first of July.* They were now *two feet and a half high;* and I was ready to begin *planting out.* The weather had been dry in the extreme. Not a drop of rain for nearly *a month.* My land was poor, but clean; and I ought to have proceeded to do the job at once. My principal man had heard so much in ridicule of the project, that he was constantly begging and praying me not to persevere. "*Every body* said it was *impossible* for the "Corn to *live!*" However, I began. I ploughed a part of the field into four-feet ridges, and, one evening, set on, thus: I put a good quantity of earth-ashes in the deep furrow between the ridges, then turned back the earth over them, and then planted the Corn on the ridge, at a foot apart. We *pulled up* the plants without ceremony, cut off their roots to half an inch long, cut off their leaves about eight inches down from their points, and, with a long setting stick, stuck them about seven inches into the ground down amongst the fresh mould and ashes.

216. This was *on the first of July* in the evening; and, not willing to be *laughed at too much,* I thought I would pause two or three

days; for, really, the sun seemed as if it would burn up the very earth. At the close of the second day, news was brought me, *that the Corn was all dead.* I went out and looked at it, and though I saw that it was not *dead,* I suffered the everlasting gloomy peal that my people rang in my ears to extort from me my consent to *the pulling up of the rest of the plants and throwing them away;* consent which was acted upon with such joy, alacrity, and zeal, that the whole lot were lying under the garden fence in a few minutes. My man intended to give them to the oxen, from the charitable desire, I suppose, of annihilating this proof of his master's folly. He would have pulled up the two rows which we had transplanted; but I would not consent to that; for, I was resolved, that they should have a *week's trial.* At the end of the week I went out and looked at them. I *slipped* out at a time *when no one was likely to see me!* At a hundred yards distance the plants looked like so many little Corn *stalks* in November; but, at twenty yards, I saw that *all was right,* and I began to reproach myself for having suffered my mind to be thwarted in its purpose by opinions opposed to principles. I saw, that the plants were all *alive,* and had begun *to shoot in the heart.* I did not stop a minute. I hastened back to the garden to see whether any of the plants, which lay in heaps, were yet alive.

217. Now, mind, the plants were put out on the first of July; the 15 succeeding days were not only *dry*, but the very *hottest* of this gloriously hot summer. The plants that had been *flung away* were, indeed, nearly all *dead;* but, some, which lay at the bottoms of the heaps, were not only *alive*, but had *shot their roots into the ground.* I resolved to plant out two rows of these, even these. While I was at it Mr. JUDGE MITCHELL called upon me. He laughed at us very heartily. This was on the *8th of July.* I challenged him *to take him three to one* my two rows against any two rows of his corn of equal length; and he is an excellent farmer on excellent land. "Then," said I, "if you are afraid to back your opinion, I do "not mind your *laugh.*"

218. On the 27th of August Mr. JUDGE MITCHELL and his brother the justly celebrated DOCTOR MITCHELL did me the honour to call here. I was gone to the mill; but they saw *the Corn.* The next day I had the pleasure to meet Doctor Mitchell, for the first time, at his brother's; and a very great pleasure it was; for a man more full of knowledge and apparently less conscious of it, I never saw in my life. But, the Corn: "What do you think of my

"Corn now?" I asked Mr. MITCHELL whether he did not think I should have won the wager. "Why, I do not know, indeed," said he, "as to "the two first planted rows."

219. On the *10th of September*, Mr. JUDGE LAWRENCE, in company with a young gentleman, saw the Corn. He examined the ears. Said that they were well-filled,, and the grains large. He made some calculations as to the amount of the crop. I think he agreed with me, that it would be at the rate of about *forty bushels to the acre.* All that now remained was to harvest the Corn, in a few weeks' time, to shell, to weigh it; and to obtain a couple of rows of equal length of every neighbour surrounding me; and then, make the comparison, the triumphant result of which I anticipated with so much certainty, that my impatience for the harvest exceeded in degree the heat of the weather, though that continued broiling hot. That very night! the night following the day when Mr. JUDGE LAWRENCE saw the Corn, eight or nine steers and heifers leaped, or broke, into my pasture from the road, kindly poked down the fence of the field to take with them four oxen of my own which had their heads tied down, and in they all went just upon the transplanted Corn, of which they left neither ear nor stem, except about two bushels

of ears which they had, in their haste, trampled under foot! What a mortification! Half an acre of fine cabbages nearly destroyed by the biting a hole in the hearts of a great part of them; turnips torn up and trampled about; a scene of destruction and waste, which, at another time, would have made me stamp and rave (if not swear) like a mad-man, seemed now nothing at all. The *Corn* was such a blow, that nothing else was felt. I was, too, both hand-tied and tongue-tied. I had nothing to wreak my vengeance on. In the case of the Boroughmongers I can repay blow with blow, and, as they have already felt, with interest and compound interest. But, there was no human being that I could blame; and, as to the depredators themselves, though in this instance, their conduct did seem worthy of another being, whom priests have chosen to furnish with horns as well as tail, what was I to do against them? In short, I had, for once in my life, to submit peaceably and quietly, and to content myself with a firm resolution never to plant, or sow, again without the protection of a fence, which an ox cannot get over and which a pig cannot go under.

220. This Corn had every disadvantage to contend with: poor land; no manure but earth-ashes burnt out of that same land; planted in

dry earth; planted in dry and hot weather; no rain to enter *two inches*, until the 8th of August, nine and thirty days after the transplanting; and yet, *every plant had one good perfect ear*, and, besides, *a small ear to each plant;* and some of the plants had *three ears*, two perfect and one imperfect. Even the *two last-planted rows*, though they were not so good, were *not bad*. My opinion is, that their produce would have been at the rate of 25 bushels to the acre; and this is not a *bad* crop of Corn.

221. For my part, if I should cultivate Corn again, I shall transplant it to a certainty. Ten days earlier, perhaps; but I shall certainly transplant what I grow. I know, that the *labour will be less*, and I believe that the crop will be far greater. No dropping the seed; no hand-hoing; no patching after the *cut-worm*, or *brown grub; no suckers;* no grass and weeds; no *stifling;* every plant has its proper space; all is clean; and one good deep ploughing, or two at most, leaves the ground as clean as a garden; that is to say, as a garden *ought* to be. The sowing of the seed in beds is one day's work (for ten acres) for one man. Hoing the young plants, another day. Transplanting, *four* dollars an acre to the very outside. "But "where are the *hands* to come from to do the "transplanting?" One would think, that, to

hear this question so often repeated, the people in America were like the Rhodian Militia, described in the beautiful poem of Dryden, "*mouths without hands.*" Far, however, is this from being the case; or else, where would the *hands* come from to do the *marking;* the *dropping and covering* of the Corn; the *hand-hoing* of it, sometimes twice; the *patching* after the grubs; the *suckering* when that work is done, as it always ought to be? Put the plague and expences of all these operations together, and you will, I believe, find them to exceed four or even six, dollars an acre, if they be all *well* done, and the Corn kept perfectly clean.

222. The transplanting of ten acres of Corn cannot be done *all in one day* by two or three men; nor is it at all necessary that it should. It may be done within the space of twelve or fourteen days. Little boys and girls, very small, will carry the plants, and if the farmer will but try, he will *stick in an acre a day himself;* for, observe, nothing is so easily done. There is no fear of *dearth.* The plants, in soft ground, might almost be poked down like so many sticks. I did not try it; but, I am pretty sure, that the *roots* might be cut all off close, so that the stump were left entire. For, mind, a *fibre*, of a stout thing, never grows again

after removal. New ones must come out of *new roots* too, or the plant, whether corn or tree, will die. When some people plant *trees*, they are so careful not to cut off the little *hairy fibres;* for these, they think, will *catch hold of the ground immediately.* If, when they have planted in the fall, they were to open the ground in June the next year, what would be their surprise to find all the hairy fibres in a mouldy state, and the new *small roots* shot out of the *big roots* of the tree, and no new fibres at all yet? for, these come out of the new small roots! It is the same with every sort of plant, except of a very small size and very quickly moved from earth to earth.

223. If any one choose to try this method of cultivating Corn, let him bear in mind, that the plants ought to be *strong*, and nearly *two feet high.* The leaves should be shortened by all means; for, they *must* perish at the tops before the new flow of sap can reach them. I have heard people say, that they *have* tried transplanting Corn very often, but have never found it to answer. But *how* have they tried it? Why, when the grub has destroyed a hill, they have taken from other hills the superabundant plants and filled up the vacancy. In the first place, they have done this when the plants were *small:* that is not my plan. Then they

have put the plants in *stale hard ground:* that is not my plan. Then they have put them into ground where prosperous neighbours had *the start of them:* that is not my plan. I am not at all surprized, that they have not found their plan to *answer;* but, that is no reason that mine should not answer. The best way will be to try *three rows* in any field, and see which method requires the least labour and produces the largest crop.

224. At any rate, the facts, which I have stated upon this subject are curious in themselves; they are useful, as they shew what we may venture to do in the removing of plants; and they shew most clearly how unfounded are the fears of those, who imagine, that Corn is *injured* by ploughing between it and breaking its roots. My plants owed their vigour and their fruit to *their removal into fresh pasture;* and, the oftener the land is ploughed between growing crops of any sort (allowing the roots to shoot between the ploughings) the better it is. I remember that LORD RANELAH showed me in 1806, in his garden at Fulham, a peach tree, which he had removed in *full bloom*, and that must have been in March, and which bore a great crop of fine fruit the same year. If a *tree* can be thus dealt with, why need we fear to transplant such things as Indian Corn?

CHAP. VI.

SWEDISH TURNIPS.

225. UPON this subject I have no great deal to add to what was said in Part I. Chap. II. There are a few things, however, that I omitted to mention, which I will mention here.

226. I sow my seed by *hand*. All *machinery* is imperfect for this purpose. The wheel of the drill meets with a sudden check; it jumps; the holes are stopped; a clogging or an improper impelling takes place; a *gap* is produced, and it can never be put to rights; and, after all, the sowing upon four feet ridges is very nearly as *quickly* performed by hand. I make the drills, or channels, to sow the seed in by means of a light *roller*, which is drawn by a horse, which rolls two ridges at a time, and which has two markers following the roller, making a drill upon the top of each ridge. This saves time; but, if the *hand* do the whole, a man will draw the drills, sow the seed, and cover an acre in a day with ease.

227. The only mischief in this case, is, that of sowing *too thick;* and this arises from the

seed being so nearly of the *colour of the earth.* To guard against this evil, I this year adopted a method which succeeded perfectly. I *wetted* the seed with water a little, I then put some *whitening* to it, and by rubbing them well together, the seed became *white* instead of *brown;* so that the man when sowing, could *see* what he was about.

228. In my directions for *transplanting turnips* I omitted to mention one very important thing; the care to be taken *not to bury the heart of the plant.* I observed how necessary it was to fix the plant *firmly in the ground;* and, as the planter is strictly charged to do this, he is apt to pay little attention to the *means* by which the object is accomplished. The thing is done easily enough, if you cram the butts of the leaves down below the surface. But, this brings the earth, with the first rain at least, over the *heart* of the plant; and then it will never *grow* at all: it will just *live;* but will never increase in size one single jot. Care, therefore, must be taken of this. The fixing is to be effected by the stick being applied to the *point* of the root; as mentioned in paragraph 85. Not to fix the plant is a great fault; but to bury the heart is a much greater; for, if this be done, the plant is sure to die.

229. My own crop of Swedish Turnips this year is far inferior to that of last in every respect. The season has been singularly unfavourable to all green and root crops. The *grass* has been barer than it was, I believe, ever known to be; and, of course, other vegetables have experienced a similar fate. Yet, I have some very good turnips; and, even with such a season, they are worth more than three times what a crop of Corn on the same land would have been. I am now (25th Nov.) giving the greens to my cow and hogs. A cow and forty stout hogs eat the greens of about twenty or thirty rods of turnips in a day. My five acres of greens will last about 25 days. I give no corn or grain of any sort to these hogs, and my English hogs are *quite fat enough for fresh pork*. I have about 25 more pigs to join these forty in a month's time: about 40 more will join those before April. My cabbages on an acre and a half of ground will carry me well on till February (unless I send my Savoys to New York), and, when the cabbages are done, I have my Swedish Turnips for March, April, May and June, with a great many to sell if I choose. I have, besides, a dozen ewes to keep on the same food, with a few wethers and lambs, for my house. In June *Early Cabbages*

come in; and then the hogs feed on them. Thus the year is brought round.

230. But, what pleases me most, as to the Swedish Turnips, is, that several of my neighbours have tried the culture, and have far surpassed me in it this year. Their land is better than mine, and they have had no Borough-villains and Bank-villains to fight against. Since my Turnips were sown, I have written great part of a Grammar and have sent twenty Registers to England, besides writing letters amounting to a reasonable volume in bulk; the whole of which has made an average of *nine pages of common print a day,* Sundays included. And, besides this, I have been *twelve days* from home, on business, and about *five* on visits. Now, whatever may have been the *quality* of the writings; whether they demanded *mind* or not, is no matter: they demanded time for the *fingers* to move in, and yet, I have not written a hundred pages *by candle-light.* A man knows not what he can do 'till he *tries.* But, then, mind, I have always been up with the cocks and hens; and I have drunk nothing but milk and water. It is a saying, that "*wine* inspires *wit;*" and that "in *wine* there is *truth.*" These sayings are the apologies of drinkers. Every thing that produces *intoxication,* though in but the slightest degree, is injurious to the *mind;* whether it

be such to the body or not, is a matter of far less consequence. My Letter to Mr. TIERNEY, on the state of the Paper-Money, has, I find, produced a great and general impression in England. The subject was of great importance, and the treating it involved much of that sort of reasoning which is the most difficult of execution. That Letter, consisting of *thirty-two full pages of print*, I wrote in one day, and that, too, on the 11th of July, the hottest day in the year. But, I never could have done this, if I had been guzzling wine, or grog, or beer, or cider, all the day. I hope the reader will excuse this digression; and, for my own part, I think nothing of the charge of *egotism*, if, by indulging in it, I produce a proof of the excellent effects of *sobriety*. It is not *drunkenness* that I cry out against: that is *beastly*, and beneath my notice. It is *drinking;* for a man may be a great *drinker*, and yet no *drunkard*. He may accustom himself to swallow, 'till his belly is a sort of tub. The Spaniards, who are a very sober people, call such a man "*a wine* "*bag*," it being the custom in that country to put wine into bags, made of *skins* or *hides*. And, indeed, *wine bag* or *grog bag* or *beer bag* is the suitable appellation.

231. To return to the Swedish Turnips, it was impossible for me to attend to them in per-

son *at all;* for, if I once *got out,* I should have *kept out.* I was very anxious about them; but much more anxious about my duty to my countrymen, who have remained so firmly attached to me, and in whose feelings and views, as to public matters, I so fully participate. I left my men to do their best, and, considering the season, they did very well. I have observed before, that I never saw my *Savoys* 'till *two months* after they were planted out in the field, and I never saw some of my Swedish Turnips 'till within these fifteen days.

232. But, as I said before, some of my neighbours have made the experiment with great success. I mentioned Mr. Dayrea's crop before, at paragraph 197. Mr. HART, at South Hampstead, has a fine piece, as my son informs me. His account is, that the field looked, in October, as fine as any that he ever saw in England. Mr. JUDGE MITCHELL has a small field that were, when I saw them, as fine as any that I ever saw in my life. He had transplanted some in the driest and hottest weather; and they were exceedingly fine, notwithstanding the singular untowardness of the season.

233. Mr. JAMES BYRD of Flushing, has, however, done the thing upon the largest scale. He sowed, in June, about two acres and a half upon ridges *thirty inches* apart. They were

very fine; and, in September, their leaves met across the intervals. On the 21st of September I saw them for the second time. The field was one body of beautiful green. The weather still very dry. I advised Mr. Byrd *to plough between them* by all means; for the roots had met long before across the interval. He observed, that the horse would *trample on the leaves.* I said, "never mind: the good done "by the plough will be ten times greater than "the injury done by the breaking of leaves." He said, that, great as his fears were, he would follow my advice. I saw the turnips again on the 8th of October, when I found, that he had *begun* the ploughing; but, that the horse *made such havock amongst the leaves,* and his *workman made such clamorous remonstrances,* that, after doing a little piece, Mr. Byrd *desisted.* These were reasons wholly insufficient to satisfy me; and at the latter, *the remonstrances of a workman,* I should have ridiculed, without a grain of mercy; only I recollected, that my men had remonstrated me (partly with sorrowful looks and shakes of the head) out of my design to transplant six acres of Indian Corn.

234. Mr. BYRD's crop was about 350 bushels to an acre. I was at his house on the 23rd of this month (November); and there I heard two things from him which I communicate with

great pleasure. The first was, that, from the time he began taking up his turnips, he began feeding his cows upon the *greens;* and, that this *doubled* the quantity *of their milk.* That the greens might last as long as possible, he put them in *small heaps*, that they might not *heat*. He took up his turnips, however, nearly a month *too early.* They grow till the *hard* frosts come. The greens are not so good till they have had *some little frost;* and, the bulb should be *ripe.* I have been now (27 Nov.) about ten days cutting off my greens. The bulbs I shall take up in about ten days hence. Those that are not consumed by that time, I shall put in small heaps in the field, and bring them away as they may be wanted.

235. The other thing stated to me by Mr. BYRD pleased me very much indeed; not only an account of its being a complete confirmation of a great principle of TULL applied to land in this climate, but on account also of the candour of Mr. BYRD, who, when he had seen the result, said, "I was wrong, friend Cobbett, in "not following thy advice." And then he went on to tell me, *that the turnips in the piece which he had ploughed after the* 21*st of September* were a crop *a fourth part greater* than those adjoining them, which remained unploughed. Thus, then, let no one be afraid of breaking

the pretty leaves that look so gay; and, how false, then must be the notion, that to plough Indian-Corn in *dry weather*, or *late*, is injurious! *Why* should it not be as beneficial to Corn as to Turnips and Cabbages?

236. Mr. BYRD transplanted with his superabundant plants, about two acres and a half. These he had not taken up on the 23rd of November. They were not so fine as the others, owing, in part, to *the hearts of many having been buried*, and to the whole having been put *too deep into* the ground. But, the ridges of both fields were *too close together*. Four feet is the distance. You cannot plough clean and deep within a smaller space without throwing the earth over the plants. But, as bulk of crop is the object, it is very hard to persuade people, that *two rows are not better than one*. Mr. JUDGE MITCHELL is a true disciple of the TULLIAN SYSTEM. His rows were four feet asunder; his ridges high; all according to rule. If I should be able to see his crop, or him, before this volume goes to the press, I will give some account of the result of his labours.

237. This year has shown me, that America is not wholly exempt from that mortal enemy of turnips, the *fly*, which mawled some of mine, and which carried off a whole piece for Mr. JUDGE LAWRENCE at Bay-side. Mr. BYRD says,

that he thinks, that to soak the seed in *fish-oil* is of use as a protection. It is very easy to *try* it; but, the best security is, pretty early sowing *thick*, and transplanting. However, this has been a *singular year*; and, even this year, the ravages of the *fly* have been, generally speaking, but trifling.

238. Another enemy has, too, made his appearance: the *caterpillar*; which came about the tenth of October. These eat the leaves; and, sometimes, they will, as in England, *eat all up*, if left alone. In Mr. BYRD's field, they were proceeding on pretty rapidly, and, therefore he took up his turnips earlier than he would have done. *Wide rows* are a great protection against these *sinecure gentry* of the fields. They attacked me on the outside of a piece joining some buck-wheat, where they had been bred. When the buckwheat was cut, they sallied out upon the turnips, and, like the spawn of real Boroughmongers, they, after eating all the leaves of the first row, went on to the second, and were thus proceeding to devour the whole. I went with my plough, ploughed a deep furrow *from* the rows of turnips, as far as the caterpillars had gone. Just shook the plants and gave the top of the ridge a bit of a sweep with a little broom. Then *burried them alive*, by turning the furrows back. Oh! that the people of

England could treat the Borough-villains and their swarms in the same way! Then might they hear without envy of the easy and happy lives of American farmers!

239. A good *sharp frost* is the only complete doctor for this complaint; but, wide rows and ploughing will do much, where the attack is made *in line,* as in my case. Sometimes, however, the enemy starts up, here and there, all over the field; and then you must plough the whole field, or be content with turnips *without greens,* and with a diminished crop of turnips into the bargain. Mr. BYRD told me, that the caterpillars did *not attack the part of the field which he ploughed after the* 21*st of September* with nearly so much fury as they attacked the rest of the field! To be sure; for, the turnip leaves there, having received fresh vigour from the ploughing, were of a taste more *acrid;* and, you always see, that insects and reptiles, that feed on leaves and bark, choose the most sickly or feeble plants to begin upon, because the juices in them are sweeter. So that here is another reason, and not a weak one, for *deep* and *late* ploughing.

240. I shall speak again of Swedish turnips when I come to treat of *hogs;* but, I will here add a few remarks on the subject of *preserving* the roots. In paragraph 106, I described the

manner in which I *stacked* my turnips last year. That did very well. But, I will not, this year, make any hole in the ground, I will pile up about thirty bushels upon the level ground, in a pyramidical form, and then, to keep the earth from running amongst them, put over a little straw, or leaves of trees, and about four or five inches of earth over the whole. For, mind, the object is not to *prevent freezing*. The turnips will freeze as hard as stones. But, so that they do not *see the sun*, or *the light*, till they are *thawed*, it is no matter. This is the case even with apples. I preserved *white turnips* this way last year. Keep the *light out*, and all will be safe with every root that I know any thing of, except that miserable thing, the *potatoe*, which, consisting of earth, of a small portion of flour, and of water *unmixed with sugar*, will freeze to perdition, if it freeze at all. Mind, it is no matter to the animals, whether the Swedish turnip, the white turnip, or the cabbage, be *frozen*, or not, at the time when they eat them. They are just as good; and are as greedily eaten. Otherwise, how would our sheep in England *fatten* on turnips (even white turnips) in the open fields and amidst snows and hard frosts? But, a potatoe, let the frost once touch it, and it is *wet dirt*.

241. I am of opinion, that if there were *no*

earth put over the turnip heaps, or stacks, it would be better; and, it would be much *more convenient*. I shall venture it for a part of my crop; and I would recommend others to try it. The *Northern Winter* is, therefore, no objection to the raising of any of these crops; and, indeed, the crops are far more necessary there than to the Southward, because the Northern Winter is so much longer than the Southern. Let the snows (even the Nova Scotia snows) come. There are the crops safe. Ten minutes brings in a waggon load at any time in winter, and the rest remain safe till spring.

242. I have been asked how I would manage the Swedish turnips, so as to keep them 'till *June* or *July*. In April (for Long Island); that is to say, when the roots begin to *shoot* out greens, or, as they will be, *yellows*, when hidden from the *light*.—Let me stop here a moment, to make a remark which this circumstance has suggested. I have said before, that if you keep the bulbs from the *light*, they will freeze and thaw without the least injury. I was able to give no *reason* for this; and who can give a *reason* for leaves being *yellow* if they grow *in the dark*, and *green*, if they grow in the *light?* It is not the *sun* (except as the *source of light*) that makes the *green;* for any plant that grows in *constant shade* will be green; while

one that grows in the *dark* will be *yellow*. When my son, JAMES, was about *three* years old, LORD COCHRANE, lying against a green bank in the garden with him, had asked him many questions about the sky, and the river, and the sun and the moon, in order to learn what were the notions, as to those objects, in the mind of a child. JAMES grew tired, for, as ROUSSEAU, in his admirable exposure of the folly of teaching *by question and answer*, observes, nobody *likes to be questioned*, and especially children. "Well," said JAMES, "now *you tell me something:* what "is it that *makes the grass green*." His Lordship told him it was the *sun*. "Why," said JAMES, pulling up some grass, "you see it is "*white down here*." "Aye," replied my Lord, "but that is because the sun cannot *get at it*." "How *get at it?*" said James: "The sun makes "it *hot* all the way down." LORD COCHRANE came in to me, very much delighted: "Here," said he, "little JEMMY has started a fine sub-"ject of dispute for all the philosophers." If this page should have the honour to meet the eye of LORD COCHRANE, it will remind him of one of the many happy hours that we have passed together, and I beg him to regard any mention of the incident as a mark of that love and respect which I bear towards him, and of the ardent desire I constantly have to see him avenged on all vile, cowardly, perjured and infamous persecutors.

243. When any one has told me, what it is that *makes* "grass green," I shall be able to tell him what it is that *makes* darkness preserve turnips; and, in the meanwhile, I am quite content with a perfect knowledge of the effects.

244. So far for the preservation *while winter lasts;* but, then, how to manage the roots when *spring* comes? Take the turnips out of the heaps; spread them upon the ground round about, or any where else in the sun. Let them get *perfectly dry*. If they lie *a month* in sun and rain alternately, it does not signify. They will take no injury. Throw them *on a barn's floor;* throw them into a *shed;* put them any where out of the way; only do not put them in *thick heaps*; for then they will heat, perhaps, and grow a little. I believe they may be kept the *whole year* perfectly sound and good; but, at any rate, I kept them thus, last year, *'till July*.

245. Of *saving seed* I have some little to say. I saved some, in order to see whether it *degenerated;* but, having, before the seed was ripe, had such complete proof of the degeneracy of *cabbage seed;* having been assured by Mr. WILLIAM SMITH, of Great Neck, that the

Swedish turnip seed had degenerated with him to a long whitish root; and, having, besides, seen the long, pale looking things in New York Market in June; I took no care of what I had growing, being *sure* of the real sort from England. However, Mr. BYRD's were from his *own seed,* which he has saved for several years. They differ from mine. They are *longer* in proportion to their circumference. The leaf is rather *more pointed,* and the inside of the bulb is not of *so deep* a yellow. Some of Mr. BYRD's have a little hole towards the crown, and the flesh is spotted with white where the green is cut off. He ascribes these defects to the season; and it may be so; but, I perceive them in none of my turnips, which are as clear and as sound, though not so large, as they were last year.

246. *Seed* is a great matter. Perhaps the best way, for farmers in general, would be always to *save some,* culling the plants carefully, as mentioned in paragraph 32. This might be sown, and also some English seed, the expense being so very trifling compared with the value of the object. At any rate, by saving some seed, a man has *something* to sow; and he has it always ready. He might change his seed once in three or four years. But, never forgetting carefully to select the plants, from which the seed is to be raised.

POSTSCRIPT TO THE CHAPTER ON SWEDISH TURNIPS.

247. Since writing the above, I have seen Mr. JUDGE MITCHELL, and having requested him to favour me with a written account of his experiment, he has obligingly complied with my request in a letter, which I here insert, together with my answer.

Ploudome, 7 Dec. 1818.

DEAR SIR,

248. About the first of June last, I received the First Part of your *Year's Residence in the United States*, which I was much pleased with, and particularly the latter part of the book, which contains a treatise on the culture of the Ruta Baga. This mode of culture was new to me, and I thought it almost impossible that a thousand bushels should be raised from one acre of ground. However, I felt very anxious to try the experiment in a small way.

249. Accordingly, on the 6th day of June, I ploughed up a small piece of ground, joining my salt meadow, containing *sixty-five rods*, that had not been ploughed for nearly thirty years. I ploughed the ground deep, and spread on it

about ten waggon loads of *composition manure;* that is to say, rich earth and yard manure mixed in a heap, a layer of each alternately. I then harrowed the ground with an iron-toothed harrow, until the surface was mellow, and the manure well mixed with the earth.

250. On the first of July I harrowed the ground over several times, and got the surface in good order; but, in consequence of such late ploughing, I dared not venture to cross-plough, for fear of tearing up the sods, which were not yet rotten. On the 7th of July I ridged the ground, throwing four furrows together, and leaving the tops of the ridges four feet asunder, and without putting in any manure. I went very shoal with the plough, because deep ploughing would have turned up the sods.

251. On the eighth of July I sowed the seed, in single rows on the tops of the ridges, on all the ridges except about eighteen. On eight of these I sowed the seed on the 19th of July, when the first sowing was up, and very severely attacked by the *flea;* and I was fearful of losing the whole of the crop by that insect. About the last of July there came a shower, which gave the turnips a start; and, on the eighth day of August I *transplanted* eight of the remaining rows, *early in the morning*. The weather was now *very dry*, and the turnips sown on the 19th

of July were just coming up. On the 10th of August I transplanted the two other rows at *mid-day*, and, in consequence of such dry weather, the *tops all died:* but, in a few days, began to look green. And, in a few weeks, those that had been transplanted looked as thrifty as those that had been sown.

252. On the 10th of August I regulated the sown rows, and left the plants standing from six to twelve inches apart.

253. A part of the seed I received from you, and a part I had from France a few years ago. When I gathered the crop, the transplanted turnips were nearly as large as those that stood where they were sown.

254. The following is the produce: *Two hundred and two bushels on sixty-five rod of ground;* a crop arising from a mode of cultivation for which, Sir, I feel very much indebted to you. This crop, as you will perceive, wants but two bushels and a fraction of *five hundred bushels to the acre;* and I verily believe, that, on his mode of cultivation, an acre of land, which will bring a hundred bushels of *corn ears*, will produce from *seven to eight hundred bushels* of the Ruta Baga Turnip.

255. Great numbers of my turnips weigh *six pounds* each. The *greens* were almost wholly destroyed by a *caterpillar*, which I never before

saw; so that I had no opportunity of trying the use of them as cattle-food; but, as to the *root*, cattle and hogs eat it greedily, and cattle as well as hogs eat up the little bits that remain attached to the fibres, when these are cut from the bulbs.

256. I am now selling these turnips at *half a dollar a bushel.*

257. With begging you to accept of my thanks for the useful information, which, in common with many others, I have received from your Treatise on this valuable plant,

I remain,

Dear Sir,

Your most obedient servant,

SINGLETON MITCHELL.

To Mr. William Cobbett,
Hyde Park.

258. P. S. I am very anxious to see the Second Part of your *Year's Residence.* When will it be published?

ANSWER.

Hyde Park, 9*th Dec.* 1818.

DEAR SIR,

259. Your letter has given me very great pleasure. You have *really tried* the thing:

you have given it a *fair* trial. Mr. TULL, when people said of his horse-hoing system, that they had tried *it*, and found it not to answer, used to reply: " *What* have they tried? all lies in " the little word IT."

260. You have really tried *it;* and very interesting your account is. It is a complete answer to all those, who talk about *loss of ground* from four-feet ridges; and especially when we compare your crop with that of Mr. JAMES BYRD, of Flushing; whose ground was prepared at an early season; who manured richly; who kept his land like a neat garden; and, in short, whose field was one of the most beautiful objects of which one can form an idea; but, whose ridges were about *two feet and a half* apart, instead of *four feet,* and who had *three hundred and fifty bushels* to the acre, while you, with all your disadvantages of late ploughing and sods beneath, had at the rate of *five hundred bushels.*

261. From so excellent a judge as you are, to hear commendation of my little Treatise, must naturally be very pleasing to me, as it is a proof that I have not enjoyed the protection of America without doing something for it in return. Your example will be followed by thousands; a new and copious source of human sustenance will be opened to a race of free and

happy people; and to have been, though in the smallest degree, instrumental in the creating of this source, will always be a subject of great satisfaction, to,

Dear Sir,

Your most obedient,

And most humble servant,

WM. COBBETT.

262. P. S. I shall to-morrow send the *Second Part of my Year's Residence* to the press. I dare say it will be ready in three weeks.

263. I conclude this chapter by observing, that a boroughmonger hireling, who was actually fed with pap, purchased by money paid to his father by the minister PITT, *for writing and publishing lies against the Prince of Wales and the Duke of York*, the acknowledgment of the facts relating to which transaction, *I saw in the father's own hand-writing*; this hireling, when he heard of my arrival on Long Island, called it my LEMNOS, which allusion will, I hope, prove not to have been wholly inapt; for, though my life is precisely the reverse of that of the unhappy PHILOCTETES, and though I do not hold the arrows of HERCULES, I do possess *arrows*; I make them felt too at a great distance, and, I am not certain, that my arrows

are not destined to be the only means of destroying the Trojan Boroughmongers.

264. Having introduced a *Judge* here by *name*, it may not be amiss to say, for the information of my English readers, what sort of persons these Long-Island Judges are. They are, some of them, *Resident* Judges, and others *Circuit* Judges. They are all gentlemen of known *independent fortune*, and of known excellent characters and understanding. They receive a mere *acknowledgment* for their services; and they are, in all respects, *liberal gentlemen*. Those with whom I have the honour to be acquainted have fine and most beautiful estates; and I am very sure, that what each actually expends in acts of *hospitality* and *benevolence* surpasses what such a man as *Burrough*, or *Richards*, or *Bailey*, or *Gibbs*, or, indeed, any of the set, expends upon every thing, except taxes. Mr. JUDGE LAURENCE, who came to invite me to his house as soon as he heard of my landing on the Island, keeps a house such as I never either saw or heard of before. My son JAMES went with a message to him a little while ago, and, as he *shot his way along*, he was in his shooting dress. He found a whole house full of company, amongst whom were the celebrated Dr. MITCHELL and Mr. CLINTON, the

Governor of this state; but, they made him stay and dine. Here was he, a boy, with his rough, shooting dress on, dining with Judges, Sheriffs, and Generals, and with the Chief Magistrate of a Commonwealth more extensive, more populous, and forty times as rich as Scotland; a Chief Magistrate of very great talents, but in whom empty pride forms no ingredient. Big wigs and long robes and supercilious airs, are necessary only when the object is to *deceive* and *overawe* the people. I'll engage that to supply Judge Laurence's house *that one week* required a greater sacrifice of animal life than merciful Gibbs's kitchen demands in a year: but, then, our hearty and liberal neighbour never deals in human sacrifices.

CHAP. VII.

POTATOES.

265. I HAVE made no experiments as to this root, and I am now about to offer my opinions as to the mode of cultivating it. But, so much has been said and written *against me* on account of my scouting the idea of this root being proper as *food for man*, I will, out of respect for public opinion, here state my *reasons* for thinking that the Potatoe is a root, *worse than useless.*

266. When I published some articles upon this subject, in England, I was attacked by the *Irish* writers with as much fury as the Newfoundlanders attack people who speak against the Pope; and with a great deal less reason; for, to attack a system, which teaches people to fill their bellies with fish for the good of their souls, might appear to be dictated by malice against the sellers of the fish; whereas, my attack upon Potatoes, was no attack upon the sons of St. Patrick, to whom, on the contrary, I wished a better sort of diet to be afforded. Nevertheless, I was told, in the Irish papers,

not that I was a *fool:* that might have been *rational;* but, when I was, by these zealous Hibernians, called a *liar*, a *slanderer*, a *viper*, and was reminded of all my *political sins*, I could not help thinking, that, to use an Irish Peeress's expression with regard to her Lord, there was a little of the Potatoe *sprouting out of their head.*

267. These rude attacks upon me even were all *nameless*, however; and, with nameless adversaries I do not like to join battle. Of one thing I am very glad; and that is, that the Irish *do not like to* live upon what their accomplished countryman DOCTOR DRENNAN, calls "Ireland's *lazy root.*" There is more sound political philosophy in that poem than in all the enormous piles of Plowden and Musgrave. When I called it a *lazy root;* when I satyrized the use of it; the Irish seemed to think, that their national *honour* was touched. But, I am happy to find, that it is not *taste*, but *necessity*, which makes them mess-mates with the pig; for when they come to this country; they invariably prefer to their "*favourite root*," not only fowls, geese, ducks and turkeys, but even the flesh of oxen, pigs and sheep!

268. In 1815, I wrote an article, which I will here insert, because it contains my opinions upon this subject. And when I have done

that, I will add some calculations as to the comparative value of an acre of wheat and an acre of potatoes. The article was a letter to the *Editor of the Agricultural Magazine*; and was in the following words.

To the Editor of the Agricultural Magazine.

Sir,

269. In an article of your Magazine for the month of September last, on the subject of my Letters to Lord Sheffield, an article with which, upon the whole, I have reason to be very proud, you express your dissent with me upon some matters, and particularly relative to *potatoes*. The passage to which I allude, is in these words: "As to a former diatribe of his "on Potatoes, we regarded it as a pleasant ex-"ample of argument for argument's sake; as "an agreeable jumble of truth and of mental "rambling."

270. Now, Sir, I do assure you, that I never was more serious in my life, than when I wrote the essay, or, rather, casually made the observations against the cultivation and use of this *worse than useless root*. If it was argument for argument's sake, no one, that I can recollect, ever did me the honour to *show* that the argu-

ment was fallacious. I think it a subject of great importance; I regard the praises of this root and the preference given to it before corn, and even some other roots, to have arisen from a sort of monkey-like imitation. It has become, of late years, the *fashion* to extol the virtues of potatoes, as it has been to admire the writings of Milton and Shakespear. God, *almighty* and all *fore-seeing*, first permitting his chief angel to be disposed to rebel against him; his permitting him to enlist whole squadrons of angels under his banners; his permitting this host to come and dispute with him the throne of heaven; his permitting the contest to be long, and, at one time, doubtful; his permitting the devils to bring cannon into this battle in the clouds; his permitting one devil or angel, I forget which, to be split down the middle, from crown to crotch, as we split a pig; his permitting the two halves, intestines and all, to go slap, up together again, and become a perfect body; his, then, causing all the devil host to be tumbled head-long down into a place called Hell, of the local situation of which no man can have an idea; his causing gates (iron gates too) to be erected to keep the devil in; his permitting him to get out, nevertheless, and to come and destroy the peace and happiness of his new creation; his causing his son to take *a pair of compasses* out of *a drawer*, to trace the form of the earth: all this, and, indeed, the whole of Milton's poem, is such barbarous trash, so outrageously offensive to reason and to common sense, that one is naturally led to wonder how it can have been tolerated by a people, amongst whom astronomy, navigation, and chemistry are understood. But, it is the *fashion* to turn up the eyes, when Paradise Lost is mentioned; and, if you fail herein you want *taste;* you want *judgment* even, if you do not admire this absurd and ridiculous stuff, when, if one of your relations were to write a letter in the same strain, you would send him to a mad-house and take his estate. It is the sacrificing of *reason* to *fashion*. And as to the other "Divine Bard," the case is still more provoking. After his ghosts, witches, sorcerers, fairies, and monsters; after his bombast and puns and smut, which appear to have been not much relished by his comparatively rude contemporaries, had had their full swing; after hundreds of thousands of pounds had been expended upon embellishing his works; after numerous commentators and engravers and painters and booksellers had got fat upon the trade; after *jubilees* had been held in honour of his memory; at a time when there were men, otherwise of apparently good sense, who were what was aptly enough termed *Shakespear-mad.* At this very moment an occurrence took place, which must have put an end, for ever, to this national folly, had it not been kept up by infatuation and obstinacy without parallel. Young IRELAND, I think his name was WILLIAM, no matter from what *motive*, though I never could see any harm in his motive, and have always thought him a man most unjustly and brutally used. No matter, however, what were the inducing circumstances, or the motives, he did write, and bring forth, as being Shakespear's, some *plays*, *a prayer*, and *a love-letter*. The learned men of England, Ireland and Scotland met to examine these performances. Some *doubted*, a few *denied;* but, the far greater part, amongst whom were Dr. PARR, Dr. WHARTON, and Mr. GEORGE CHALMERS, declared, in the most positive terms, that *no man but Shakespear* could have written those things. There was a *division;* but this division arose more from a suspicion of some trick, than from any thing to be urged against the merit of the writings. The plays went so far as to be ACTED. Long lists of subscribers appeared to the work. And, in short, it was decided, in the most unequivocal manner, that this young man of sixteen years of age had written so *nearly like Shakespear*, that a majority of the learned and critical classes of the nation most firmly believed the writings to be Shakespear's; and, there cannot be a doubt, that, if Mr. Ireland had been able to keep his secret, they would have passed for Shakespear's 'till the time shall come when the whole heap of trash will, by the natural good sense of the nation, be consigned to everlasting oblivion; and, indeed, as folly ever doats on a darling, it is very likely, that these last found productions of "*our immortal bard*" would have been regarded as his *best*. Yet, in spite of all this; in spite of what one would have thought was sufficient to make blind people see, the fashion has been kept up; and, what excites something *more* than ridicule and contempt, Mr. Ireland, whose writings had been taken for Shakespear's, was, when he *made the discovery*, treated as an impostor and a *cheat*, and hunted down with as much rancour as if he had written against the buying and selling of seats in Parliament. The *learned* men; the *sage critics;* the *Shakespear-mad folks;* were all so *ashamed*, that they endeavoured to draw the public attention from themselves to the young man. It was of *his impositions* that they now talked, and not of their *own folly*. When the witty clown, mentioned in Don Quixote, put the nuncio's audience to shame by pulling the *real pig* out from under his cloak, we do not find that that audience were, like our *learned* men, so unjust as to pursue him with reproaches and with every act that a vindictive mind can suggest. They perceived how foolish they had been, they hung down their heads in silence, and, I dare say, would not easily be led to admire the mountebank again.

271. It is *fashion*, Sir, to which in these most striking instances, sense and reason have yielded; and it is to *fashion* that the potatoe owes its general cultivation and use. If you ask me whether fashion can possibly make *a nation* prefer one sort of *diet* to another, I ask you what it is that can make *a nation* admire Shakespear? What is it that can make them call him a "Divine Bard," nine-tenths of whose works are made up of such trash as no decent man, now-a-days, would not be ashamed, and even afraid, to put his name to? What can make an audience in London sit and hear, and even applaud, under the name of Shakespear, what they would hoot off the stage in a moment, if it came forth under any other name? When folly has once given the fashion she is a very persevering dame. An American writer, whose name is GEORGE DORSEY, I believe, and who has recently published a pamphlet, called, "The "UNITED STATES AND ENGLAND, &c." being a

reply to an attack on the morals and government and learning of the Americans, in the "Quarterly Review," states, as matter of *justification*, that the People of America sigh *with delight* to see the plays of Shakespear, whom they claim as *their countryman;* an honour, if it be disputed, of which I will make any of them a voluntary surrender of my share. Now, Sir, what can induce the American to sit and hear with delight the dialogues of Falstaff and Poins, and Dame Quickely and Doll Tearsheet? What can restrain them from pelting Parson Hugh, Justice Shallow, Bardolph, and the whole crew off the stage? What can make them endure a ghost *cap-à-pie*, a prince, who, for *justice* sake, pursues his uncle and his mother, and who stabs an old gentleman in sport, and cries out "dead for a ducat! dead!" What can they find to "delight" them in punning clowns, in ranting heroes, in sorcerers, ghosts, witches, fairies, monsters, sooth-sayers, dreamers; in incidents out of nature, in scenes most unnecessarily bloody. How they must be delighted at the story of Lear putting the question to his daughters of *which loved him most*, and then dividing his kingdom among them, *according to their professions of love;* how delighted to see the fantastical disguise of Edgar, the *treading out* Gloucester's eyes, and the trick by which it is pretended he was made to believe, that he had actually fallen from the top of the cliff! How they must be delighted to see the stage filled with green boughs, like a coppice, as in Macbeth, or streaming like a slaughter-house, as in Titus Andronicus! How the young girls in America must be tickled with delight at the dialogues in Troilus and Cressida, and more especially at the pretty observations of the *Nurse*, I think it is, in Romeo and Juliet! But, it is the same all through the work. I know of one other, and *only one other*, book, so *obscene* as this; and, if I were to judge from the high favour in which these two books seem to stand, I should conclude, that wild and improbable fiction, bad principles of morality and politicks, obscurity in meaning, bombastical language, forced jokes, puns, and smut, were fitted to the minds of the people. But I do not thus judge. It is *fashion.* These books are in fashion. Every one is ashamed not to be in the fashion. It is the fashion to extol potatoes, and to eat potatoes. Every one joins in extolling potatoes, and all the world like potatoes, or pretend to like them, which is the same thing in effect.

272. In those memorable years of wisdom, 1800 and 1801, you can remember, I dare say, the grave discussions in Parliament about pota-

toes. It was proposed by some one to make *a law* to encourage the growth of them; and, if the Bill did not pass, it was, I believe, owing to the ridicule which Mr. Horne Tooke threw upon that whole system of petty legislation. Will it be believed, in another century, that the law-givers of a great nation actually passed a law to compel people to eat pollard in their bread, and that, too, not for the purpose of *degrading* or *punishing*, but for the purpose of doing the said people good by *adding* to *the quantity of bread* in a time of scarcity? Will this be believed? In every bushel of wheat there is a certain proportion of *flour*, suited to the appetite and the stomach of man; and a certain proportion of *pollard* and *bran*, suited to the appetite and stomach of pigs, cows, and sheep. But the parliament of the years of wisdom wished to cram the *whole* down the throat of man, together with the flour of other grain. And what was to become of the pigs, cows, and sheep? Whence were the pork, butter, and mutton to come? And were not these articles of human food as well as bread? The truth is, that pollard, bran, and the coarser kinds of grain, when given to cattle, make these cattle fat; but when eaten by man make him lean and weak. And yet this bill actually became a law!

273. That period of wisdom was also the period of the potatoe-mania. *Bulk* was the only thing sought after; and, it is a real fact, that Pitt did suggest the making of *beer* out of *straw*. Bulk was all that was looked after. If the scarcity had continued a year longer, I should not have been at all surprized, if it had been proposed to feed the people at rack and manger. But, the *Potatoe!* Oh! What a blessing to man! LORD GRENVILLE, at a birth-day dinner given to the foreign ambassadors, used not a morsel of bread, but, instead of it, little *potatoe cakes,* though he had, I dare say, a plenty of lamb, poultry, pig, &c. All of which had been fatted upon corn or meal, in whole or in part. Yes, Sir, potatoes will do very well along with plenty of animal food, which has been *fatted on something better* than potatoes. But, when you and I talk of the use of them, we must consider them in a very different light.

274. The notion is, that potatoes are *cheaper* than wheat *flour.* This word *cheap* is not quite expressive enough, but it will do for our present purpose. I shall consider the *cost* of potatoes, in a family, compared with that of flour. It will be best to take the simple case of the labouring man.

275. The price of a bushel of fine flour, at Botley, is, at this time, 10*s*. The weight is 56 lbs. The price of a bushel of potatoes is 2*s*. 6*d*. They are just now dug up, and are at the cheapest. A bushel of potatoes which are measured by a large bushel, weighs about 60 lbs. dirt and all, for they are sold unwashed. Allow 4 lbs. for dirt, and the weights are equal. Well, then, here is toiling Dick with his four bushels of potatoes, and John with his bushel of flour. But, to be fair, I must allow, that the relative price is not always so much in favour of flour. Yet, I think you will agree with me, that upon an average, five bushels of potatoes do cost as much as one bushel of flour. You know very well, that potatoes in London, sell for 1*d*. and sometimes for 2*d*. a pound; that is to say, sometimes for 1*l*. 7*s*. 6*d*. and sometimes for 2*l*. 15*s*. the five bushels. This is notorious. Every reader knows it. And did you ever hear of a bushel of flour selling for 2*l*. 15*s*. Monstrous to think of! And yet the tradesman's wife, looking *narrowly* to every halfpenny, trudges away to the potatoe shop to get five or six pounds of this wretched root for the purpose of *saving flour!* She goes and gives 10*d*. for ten pounds of potatoes, when she might buy five pounds of flour with the same money!

Before her potatoes come to the table, they are, even in *bulk*, less than 5 lbs. or even 3 lbs. of flour made into a pudding. Try the experiment yourself Sir, and you will soon be able to appreciate the *economy* of this dame.

276. But, to return to Dick and John; the former has got his five bushels of potatoes, and the latter his bushel of flour. I shall, by and by, have to observe upon the *stock* that Dick must lay in, and upon the stowage that he must have; but, at present, we will trace these two commodities in their way to the mouth and in their effects upon those who eat them. Dick has got five bushels at once, because he could have them a little cheaper. John may have his *Peck* or *Gallon* of flour: for that has a fixed and indiscriminating price. It requires no trick in dealing, no judgment, as in the case of the roots, which may be *wet*, or *hollow*, or *hot*; flour may be sent for by any child able to carry the quantity wanted. However, reckoning Dick's trouble and time nothing in getting home his five bushels of potatoes, and supposing him to have got the *right* sort, a "*fine* sort," which he can hardly fail of, indeed, since the whole nation is now full of "fine sort," let us now see how he goes to work to consume them. He has a piece of bacon upon the rack, but he must have some potatoes too. On goes the *pot*, but

there it may as well hang, for we shall find it in continual requisition. For this time the meat and roots boil together. But, what is Dick to have for supper? Bread? No. He shall not have bread, unless he will have bread for dinner. Put on the Pot again for supper. Up an hour before day light and on with the pot. Fill your luncheon-bag, Dick: nothing is so relishing and so strengthening out in the harvest-field, or ploughing on a bleak hill in winter, as a cold potatoe. But, be sure, Dick, to wrap your bag well up in your clothes, during winter, or, when you come to lunch, you may, to your great surprise, find your food transformed into pebbles. Home goes merry Dick, and on goes the pot again. Thus 1095 times in the year Dick's pot must boil. This is, at least, a thousand times oftener than with a bread and meat diet. Once a week baking and once a week boiling, is as much as a farm house used to require. There must be some fuel consumed in winter for warmth. But here are, at the least, 500 fires to be made for the sake of these potatoes, and, at a penny a fire, the amount is more than would purchase four bushels of flour, which would make 288 lbs. of bread, which at 7 lbs. of bread a day, would keep John's family in bread for 41 days out of the 365. This I state as a fact challenging contradiction, that, exclusive of the extra *labour*, occasioned by the cookery of potatoes, the *fuel* required, in a year, for a bread diet, would cost, in any part of the kingdom, more than would keep a family, even in baker's bread for 41 days in the year, at the rate of 71 lbs. of bread a day.

277. John, on the contrary, lies and sleeps on Sunday morning 'till about 7 o'clock. He then gets a bit of bread and meat, or cheese, if he has either. The mill gives him his bushel of flour in a few minutes. His wife has baked during the week. He has a pudding on Sunday, and another batch of bread, before the next Sunday. The moment he is up, he is off to his stable, or the field, or the coppice. His breakfast and luncheon are in his bag. In spite of frost he finds them safe and sound. They give him heart, and enable him to go through the day. His 56 lbs. of flour, with the aid of 2*d*. in yeast, bring him 72 lbs. of bread; while, after the dirt and peelings and waste are deducted, it is very doubtful whether Dick's 300 lbs. of potatoes bring 200 lbs. of even this watery diet to his lips. It is notorious, that in a pound of clean potatoes there are 11 ounces of water, half an ounce of earthy matter, an ounce of *fibrous* and *strawey* stuff, and I know not what besides. The *water* can do Dick no good, but he must swallow these 11 ounces of

water in every pound of potatoes. How far *earth* and *straw* may tend to fatten or strengthen cunning Dick, I do not know; but, at any rate, it is certain, that, while he is eating as much of potatoe as is equal in nutriment to 1 lb. of bread, he must swallow about 14 oz. of water, earth, straw, &c. for, down they must go altogether, like the Parliament's bread in the years of wisdom, 1800 and 1801. But, suppose every pound of potatoes to bring into Dick's stomach a 6th part in nutritious matter, including in the gross pound all the dirt, eyes, peeling, and other inevitable waste. Divide his gross 300 lbs. by 6, and you will find him with 50 lbs. of nutritious matter for the same sum that John has laid out in 72 lbs. of nutritious matter, besides the price of 288 lbs. of bread in a year, which Dick lays out in extra fuel for the eternal boilings of his pot. Is it any wonder that his cheeks are like two bits of loose leather, while he is pot-bellied, and weak as a cat? In order to get half a pound of nutritious matter into him, he must swallow about 50 ounces of water, earth, and straw. Without ruminating faculties how is he to bear this cramming?

278. But, Dick's disadvantages do not stop here. He must lay in his store at the beginning of winter, or he must buy through the nose. And, where is he to find *stowage?* He has no

caves. He may *pie* them in the garden, if he has none; but, he must not open the pie in frosty weather. It is a fact not to be disputed, that a full *tenth* of the potatoe crop is destroyed, upon an average of years, by the frost. His wife, or stout daughter, cannot go out to work to help to earn the means of buying potatoes. She must stay at home to *boil the pot*, the everlasting pot! There is no such thing as *a cold dinner*. No such thing as women sitting down on a hay-cock, or a shock of wheat, to their dinner, ready to jump up at the approach of the shower. Home they must tramp, if it be three miles, to the fire that ceaseth not, and the pot as black as Satan. No wonder, that in the brightest and busiest seasons of the year, you see from every cottage door, staring out at you, as you pass, a smoky-capped, greasy-heeled woman. The pot, which keeps her at home, also gives her the colour of the chimney, while long inactivity swells her heels.

279. Now, Sir, I am quite serious in these my reasons against the use of this root, as food for man. As food for other animals, in proportion to its cost, I know it to be the *worst of all roots* that I know any thing of; but, that is another question. I have here been speaking of it as food for man; and, if it be more expensive than flour to the labourer *in the country*,

who, at any rate, can stow it in pies, what must it be to tradesman's and artizan's families in *towns*, who can lay in no store, and who must buy by the ten pound or quarter of a hundred at a time? When broad-faced Mrs. Wilkins tells Mrs. Tomkins, that, so that she has "*a* "*potatoe*" for her dinner, *she does not care a farthing for bread*, I only laugh, knowing that she will twist down a half pound of *beef* with her "potatoe," and has twisted down half a pound of buttered toast in the morning, and means to do the same at tea time without prejudice to her supper and grog. But when Mrs. Tomkins gravely answers, "yes, Ma'am, there is nothing "like a potatoe; it is such a *saving* in a family," I really should not be very much out of humour to see the tête-à-tête broken up by the application of a broom-stick.

280. However, Sir, I am talking to *you* now, and, as I am not aware that there can be any impropriety in it, I now call upon you to show, that I am really wrong in my notions upon this subject; and this, I think you are, in some sort bound to do, seeing that you have, in a public manner, condemned them.

281. But, there remains a very important part of the subject yet undiscussed. For, though you should be satisfied, that 300 lbs. of potatoes are not, taking every thing into consi-

deration, more than equal to about 30 lbs. of flour, you may be of opinion, that the disproportion in the bulk of the *crops* is, in favour of potatoes, more than sufficient to compensate for this. I think this is already clearly enough settled by the *relative prices* of the contending commodities; for, if the quantity of produce was on the side of potatoes, their *price* would be in proportion.

282. I have *heard* of enormous crops of potatoes; as high, I believe, as 10 tons grow upon an acre. I have heard of 14 sacks of wheat upon an acre. I never saw above 10 grow upon an acre. The average crop of wheat is about 24 bushels, in this part of England, and the average crop of potatoes about 6 tons. The weight of the wheat 1,440 lbs. and that of the potatoes 13,440 lbs. Now, then, if I am right in what has been said above, this *bulk* of potatoes barely keeps place with that of the wheat; for, if a bushel of wheat does not make 56 lbs. of *flour*, it weighs 60 lbs. and leaves pollard and bran to make up the deficiency. Then, as to the *cost:* the ground must be equally good. The seed is equally expensive. But the potatoes must be cultivated *during their growth.* The expense of digging and cartage and stowage is not less than 2*l.* an acre at present prices. The expense of reaping,

housing, and threshing is, at present prices, 10*s*. less. The potatoes leave *no straw*, the wheat leaves straw, stubble, and gleanings for pigs. The straw is worth, at least, 3*l*. an acre, at present prices. It is, besides, *absolutely necessary*. It litters, in conjunction with other straw, all sorts of cattle; it sometimes helps to feed them; it covers half the buildings in the kingdom; and makes no small part of the people's beds. The potatoe is a robber in all manner of ways. It largely takes from the farm-yard, and returns little, or nothing to it; it robs the land more than any other plant or root, it robs the eaters of their time, their fuel, and their health; and, I agree fully with MONSIEUR TISSOT, that it robs them of their *mental powers*.

283. I do not deny, that it is a pleasant enough thing to assist in sending down lusty Mrs. Wilkins's good half-pound of fat roast-beef. Two or three ounces of water, earth, and straw, can do *her* no harm; but, when I see a poor, little, pale-faced, life-less, pot-bellied boy peeping out at a cottage door, where I ought to meet with health and vigour, I cannot help cursing the fashion, which has given such general use to this root, as food for man. However, I must say, that the chief ground of my antipathy to this root is, that it tends to *debase the common people*, as every thing does, which

brings their mode of living to be nearer that of cattle. The man and his pig, in the potatoe system, live pretty much upon the same diet, and eat nearly in the same manner, and out of nearly the same utensil. The same eternally-boiling pot cooks their common mess. Man, being master, sits at the first table; but, if his fellow-feeder comes after him, he will not *fatten*, though he will *live* upon the same diet. Mr. CURWEN found potatoes to supply the place of *hay*, being first *well cooked;* but, they did not supply the place of oats; and yet fashion has made people believe, that they are capable of supplying the place of *bread!* It is notorious, that *nothing* will *fatten* on potatoes alone. Carrots, parsnips, cabbages, will, in time, fatten sheep and oxen, and, some of them, pigs; but, upon potatoes *alone*, no animal that I ever heard of will fatten. And yet, the greater part, and, indeed, all the other roots and plants here mentioned, will yield, upon ground of the same quality, three or four times as heavy a crop as potatoes, and will, too, for a long while, set the frosts at defiance.

284. If, Sir, you do me the honour to read this letter, I shall have taken up a good deal of your time; but the subject is one of much importance in rural economy, and therefore, can-

not be wholly uninteresting to you. I will not assume the sham modesty to suppose, that my manner of treating it makes me unworthy of an answer; and, I must confess, that I shall be disappointed unless you make a serious attempt to *prove* to me, that I am in error.

I am, Sir,
Your most obedient,
And most humble Servant,
WM. COBBETT.

---

285. Now, observe, I never received any *answer* to this. Much *abuse*. New torrents of *abuse;* and, in language still more venomous than the former; for *now* the Milton and Shakespear men, the critical *Parsons,* took up the pen; and, when you have an angry *Priest* for adversary, it is not the common viper, but the rattle-snake that you have to guard against. However, as no one put his *name* to what he wrote, my remarks went on producing their effect; and a very considerable effect they had.

286. About the same time Mr. TIMOTHY BROWN of Peckham Lodge, who is one of the most *understanding* and most worthy men I ever had the honour to be acquainted with, furnished me with the following comparative estimate relative to *wheat* and *potatoes.*

PRODUCE OF AN ACRE OF WHEAT.

287. Forty bushels is a *good* crop; but from fifty to sixty may be grown.

		Pounds of Wheat.
40 bushels 60 pounds a bushel . .		2,400
$45\frac{1}{2}$ pounds of flour to each bushel of wheat	1820	
13 pounds of offal to each bushel	520	
Waste	60	
		2,400
The worth of offal is about that of one bushel of flour; and the worth of straw, 2 tons, each worth 2*l.* is equal to six bushels of flour	$318\frac{1}{2}$	
		Pounds of Flour.
So that the total yield, *in flour*, is .		2,139
		Pounds of Bread.
Which *will* make of *bread*, at the rate of 9 pounds of bread from 7 pounds of flour		$2,739\frac{1}{2}$

PRODUCE OF AN ACRE OF POTATOES.

288. Seven tons, or 350 bushels, is a *good* crop; but ten tons, or 500 bushels *may* be grown.

	Pounds of Potatoes.
Ten tons, or	22,400

	Pounds of Flour.
Ten pounds of Potatoes contain one pound of flour	2,240

	Pounds of Bread.
Which *would*, if it *were possible* to extract the flour and get it in a dry state, make of bread . .	2,880

289. Thus, then, the *nutritious contents* of the Potatoes surpasses that of the wheat but by a few pounds; but to get at those contents, unaccompanied with *nine times their weight* in earth, straw, and water, is *impossible.* Nine pounds of earth, straw and water must, then, be swallowed, in order to get at the one pound of flour!

290. I beg to be understood as saying nothing against the *cultivation* of potatoes in any place, or near any place where there are people willing to consume them at *half a dollar*

a bushel, when wheat is *two dollars a bushel.* If any one will buy *dirt* to eat, and if one can get dirt to him with more profit than one can get wheat to him, let us supply him with dirt by all means. It is his *taste* to eat dirt; and, if his taste have nothing immoral in it, let him, in the name of all that is ridiculous, follow his taste. I know *a prime Minister,* who picks his nose and regales himself with the contents. I solemnly declare this to be true. I have witnessed the worse than beastly act scores of times; and yet I do not know, that he is much more of a beast than the greater part of his associates. Yet, if this were *all;* if he were chargeable with nothing but this; if he would confine his *swallow* to this, I do not know that the nation would have any right to interfere between his nostrils and his gullet.

291. Nor do I say, that it is *filthy* to eat potatoes. I do not ridicule the using of them as *sauce.* What I laugh at is, the idea of the use of them being a *saving;* of their *going further* than bread; of the cultivation of them in lieu of wheat *adding to the human sustenance of a country.* This is what I laugh at; and laugh I must as long as I have the above estimate before me.

292. As food for cattle, sheep or hogs, this

is the *worst* of all the green and root crops; but, of this I have said enough before; and, therefore, I now dismiss the Potatoe with the hope, that I shall never again have to write the word, or to see the thing.

CHAP. VIII.

COWS, SHEEP, HOGS, AND POULTRY.

293 *Cows.*—With respect to cows, need we any other facts than those of Mr. BYRD to prove how advantageous the Swedish turnip culture must be to those who keep cows in order to make butter and cheese. The *greens* come to supply the place of grass, and to add a *month* to the feeding on green food. They come just at the time when cows, in this country, are *let go dry*. It is too hard work to squeeze butter out of straw and corn stalks; and, if you could get it out, it would not, pound for pound, be nearly so good as *lard*, though it would be full as white. To give cows *fine hay* no man thinks of; and, therefore, dry they must be from November until March, though a good piece of cabbages added to the turnip greens would keep them on in milk to their calving time; or, 'till within a month of it at any rate. The bulbs of Swedish turnips are *too valuable* to give to cows; but the cabbages, which are so easily raised, may be made subservient to their use.

294. *Sheep.*—In the *First Part* I have said how I fed my sheep upon Swedish turnips. I have now only to add, that, in the case of *early lambs for market*, cabbages, and especially *savoys*, in February and March, would be excellent for *the ewes.* Sheep love *green.* In a turnip field, they never touch the bulb, till every bit of green is eaten. I would, therefore, for this purpose, have some cabbages, and, if possible, of the *savoy* kind.

295. *Hogs.*—This is the main object, when we talk of raising green and root crops, no matter how near to or how far from the spot where the produce of the farm is to be consumed. For, pound for pound, the hog is the most valuable animal; and, whether fresh or salted, is the most easily conveyed. Swedish turnips or cabbages or Mangel Wurzel will *fatten* an *ox;* but, that which would, in four or five months fatten the ox, would keep fifteen *August Pigs* from the grass going to the grass coming, on Long Island. Look at *their worth in June*, and compare it with the few dollars that you have got by fatting the ox; and look also at the *manure* in the two cases. A farmer, on this Island fatted two oxen last winter upon corn. He told me, after he had sold them, that, if he had *given the oxen away*, and *sold the corn*, he should have had more money

in his pocket. But, if he had kept, through the winter, four or five summer pigs upon this corn, would *they* have eaten all his corn to no purpose? I am aware, that pigs get something at an ox-stable door; but, what a process is this!

296. My hogs are now *living wholly* upon *Swedish turnip greens,* and, though I have taken no particular pains about the matter, they look very well, and, for store hogs and sows, are as fat as I wish them to be. My English hogs are sleek, and fit for *fresh pork;* and *all* the hogs not only eat the greens but do well upon them. But, observe, I give them *plenty three times a day.* In the forenoon we get a good waggon load, and that is for three meals. This is a main thing, this *plenty;* and, the farmer must see to it with his OWN EYES; for, workmen are all *starvers,* except of themselves. I never had a man in my life, who would not starve a hog, if I would let him; that is to say, if the food was to be got by some labour. You must, therefore, see to *this;* or, you do not *try* the thing at all.

297. Turnip *greens* are, however, by no means equal to cabbages, or even to *cabbage leaves.* The cabbage, and even the leaf, is the *fruit* of the plant; which is not the case with the Turnip green. Therefore the latter must,

especially when they follow summer cabbages, be given in greater proportionate quantities.

298. As to the *bulb* of the Swedish turnip, I have said enough, in the First Part, as food for hogs; and I should not have mentioned the matter again, had I not been visited by two gentlemen, who *came on purpose* (from a great distance) to *see*, whether hogs *really* would eat Swedish turnips! Let not the English farmers *laugh* at this; let them not imagine, that the American farmers are a set of simpletons on this account: for, only about thirty years ago, the English farmers would, not, indeed, have gone a great distance to ascertain the fact, but would have said at once, that the *thing was false*. It is not more than about four hundred years since the Londoners were wholly supplied with cabbages, spinage, turnips, carrots, and all sorts of garden stuff *from Flanders*. And now, I suppose, that one single parish in Kent grows more garden stuff than all Flanders. The first settlers came to America long and long before even the *white turnip* made its appearance in the *fields* in England. The successors of the first settlers trod in the foot-steps of their fathers. The communication with England did not bring out *good English farmers*. Books made little impression unaccompanied with ac-

tual experiments on the spot. It was reserved for the Boroughmongers, armed with gags, halters, and axes, to drive from England experience and public spirit sufficient to introduce the culture of the green and root crops to the fields of America.

299. The first gentleman, who came to see whether hogs would eat Swedish turnips saw some turnips tossed down on the grass to the hogs, which were eating sweet little loaved cabbages. However, they eat the turnips too before they left off. The second, who came on the afternoon of the same day, saw the hogs eat some bulbs chopped up. The hogs were pretty hungry, and the quantity of turnips small, and there was such a shoving and pushing about amongst the hogs to snap up the bits, that the gentleman observed, that they "*liked them as* "*well as corn.*"

300. In paragraph 134 I related a fact of a neighbour of mine in Hampshire having given his Swedish turnips, *after they had borne seed*, to some lean pigs, and had, with that food, made them fit for *fresh pork*, and sold them as such. A gentleman from South Carolina was here in July last, and I brought some of mine which had then *borne seed*. They were *perfectly sound.* The hogs ate them as well as if they had not borne seed. We boiled some in the kitchen for

dinner; and they appeared as good as those eaten in the winter. This shews clearly how well this root *keeps*.

301. Now, these facts being, I hope, undoubted, is it not surprising, that, in many parts of this fine country, it is the rule to keep only *one pig for every cow!* The cow seems as necessary to the pig as the pig's mouth is necessary to his carcass. There are, for instance, six cows; therefore, when they begin to give milk in the spring, six pigs are set on upon the milk, which is given them with a suitable proportion of pot liquor (a *meat* pot) and of rye, or Indian, meal, making a diet far superior to that of the families of labouring men in England. Thus the pigs go on 'till the time when the cows (for want of moist food) become dry. Then the pigs are shut up, and have the new sweet Indian corn heaped into their stye till they are quite fat, being half fat, mind, all the summer long, as they run barking and capering about. Sometimes they turn sulky, however, and will not eat enough of the corn; and well they may, seeing that they are deprived of their *milk*. Take a child from its *pap* all at once, and you will find, that it will not, for a long while, relish its new diet. What a system! but if it must be persevered in, there might, it appears to me, be a great improvement made even in it; for, the

labour of milking and of the subsequent operations, all being performed by *women*, is of great inconvenience. Better let each pig suck its adopted mother at once, which would save a monstrous deal of labour, and prevent all possibility of waste. There would be no *slopping* about; and, which is a prime consideration in a dairy system, there would be *clean milking*; for, it has been proved by DOCTOR ANDERSON, that the last drop is *fourteen times* as good as the first drop; and, I will engage, that the grunting child of the lowing mother would *have that last drop* twenty times a day, or would pull the udder from her body. I can imagine but one difficulty that can present itself to the mind of any one disposed to adopt this improvement; and that is, the teaching of the pig to suck the cow. This will appear a difficulty to those only who think unjustly of the understandings of pigs: and, for their encouragement, I beg leave to refer them to DANIEL'S RURAL SPORTS, where they will find, that, in Hampshire, Sir John Mildmay's gamekeeper, Toomer, taught a sow to point at partridges and other game; to quarter her ground like a pointer, to back the pointers, when she hunted with them, and to be, in all respects, the most docile pointer of the finest nose. This fact is true beyond all doubt. It is known to many men now alive. Judge, then, how easily a pig might be taught to milk a cow, and what a "*saving of labour*" this would produce!

302. It is strange what comfort men derive even from the deceptions which they practice upon themselves. The milk and fat pot-liquor and meal are, when put together, called, in Long Island, *swill*. The *word* comes from the farm-houses in England, but it has a new *meaning* attached to it. There it means the mere *wash;* the mere *drink* given to store hogs. But, here it means *rich fatting food.* "There, friend "Cobbett," said a gentleman to me, as we looked at his pigs, in September last, "do thy English "pigs look better than these?" "No," said I, "but what do these live on?" He said he had given them all summer, "*nothing* but "*swill.*" "Aye," said I, "but *what is* swill?" It was, for *six pigs*, nothing at all, *except* the milk of *six very fine cows*, with a bin of *shorts and meal* always in requisition, and with the daily supply of liquor from a pot and a spit, that boils and turns without counting the cost.

303. This is very well for those who do not care a straw, whether their pork cost them seven cents a pound or half a dollar a pound; and, I like to see even the *waste;* because it is a proof of the easy and happy life of the farmer. But, when we are talking of *profitable* agricul-

ture, we must examine this *swill* tub, and see what it contains. To keep pigs to a profit, you must carry them on *to their fatting time* at little expence. Milk comes from all the grass you grow and almost the whole of the dry fodder. Five or six cows will sweep a pretty good farm as clean as the turnpike road. Pigs, till *well weaned* must be kept upon *good food.* My pigs will always be fit to go out of the weaning stye at *three months* old. The common pigs require *four months.* Then out they go never to be fed again, except on grass, greens, or roots, till they arrive at the age to be fattened. If they will not keep themselves in *growing order* upon this food, it is better to shoot them at once. But, I never yet saw a hog that would not. The difference between the good sort and the bad sort, is, that the former will always be fat enough for *fresh* pork, and the latter will not; and that, in the fatting, the former will not require (weight for weight of animal) more than *half* the food that the latter will to make them equally fat.

304. Out of the milk and meal system another monstrous evil arises. It is seldom that the hogs come to a *proper age* before they are killed. A hog has not got his growth till he is full *two years old.* But, who will, or can, have the patience to see a hog *eating* Long-Island

swill for two years? When a hog is only 15 or 16 months old, he will lay on two pounds of fat for every one pound that will, out of the same quantity of food, be laid on by an eight or ten months' pig. Is it not thus with every animal? A stout boy will be like a herring upon the very food that would make his father fat, or kill him. However, this fact is too notorious to be insisted on.

305. Then, the young meat is not so nutritious as the old. Steer-beef is not nearly so good as ox-beef. Young wether mutton bears the same proportion of inferiority to old wether mutton. And, what reason is there, that the principle should not hold good as to hog-meat? In Westphalia, where the fine hams are made, the hogs are never killed under *three years old.* In France, where I saw the fattest pork I ever saw, they keep their fatting hogs to the same age. In France and Germany, the people do not eat the hog, *as hog:* they use the hog *to put fat into other sorts of meat.* They make holes in beef, mutton, veal, turkeys and fowls, and, with a tin tube, draw in bits of fat hog, which they call *lard,* and, as it is *all fat,* hence comes it that we call the inside fat of a hog, *lard.* Their beef and mutton and veal would be very poor stuff without the aid of the hog; but, with that aid, they make them all exceed-

ingly good. Hence it is, that they are induced to keep their hogs till they have *quite done growing;* and, though their sort of hogs *is the very worst* I ever saw, their hog meat was *the very fattest.* The common weight in Normandy and Brittany is *from six to eight hundred pounds.* But, the poor fellows there do not slaughter away as the farmers do here, ten or a dozen hogs at a time, so that the sight makes one wonder whence are to come the mouths to eat the meat. In France *du lard* is a thing to *smell to,* not to *eat.* I like the eating far better than the smelling system; but when we are talking about farming for *gain,* we ought to inquire how any given weight of meat can be obtained at the *cheapest rate.* A hog in his third year, would, on the American plan, suck half a dairy of cows perhaps; but, then, mind, he would, *upon a third part of the fatting food,* weigh down four Long Island "*shuts,*" the average weight of which is about *one hundred and fifty pounds.*

306. A hog, upon rich food, will be much bigger *at the end of a year,* than a hog upon *good growing diet;* but, he will not be bigger at the end of *two years,* and especially at the end of *three years.* His *size* is not to be *forced on,* any more than that of a child, beyond a certain point.

307. For these reasons, if I were settled as a farmer, I would let my hogs have *time to come to their size.* Some sorts come to it at an earlier period, and this is amongst the good qualities of my English hogs; but, to do the thing well, even they ought to have *two years* to grow in.

308. The reader will think, that I shall never cease talking about *hogs;* but, I have now done, only I will add, that, in keeping hogs in *a growing state,* we must never forget their *lodging!* A few boards, flung carelessly over a couple of rails, and no litter beneath, is not the sort of bed for a hog. A place of suitable size, large rather than small, well sheltered on every side, covered with a roof that lets in no wet or snow. No opening, except a door-way big enough for a hog to go in; and the floor constantly well bedded with leaves of trees, *dry,* or, which is the best thing, and what a hog deserves, *plenty of clean straw.* When I make up my hogs' lodging place for winter, I look well at it, and consider, whether, upon a pinch, I could, for once and away, make shift to lodge in it myself. If I *shiver at the thought,* the place is not good enough for my hogs. It is not in the nature of a hog to sleep in the cold. Look at them. You will see them, if they have the means, *cover themselves over* for the night. This is what is

done by neither horse, cow, sheep, dog nor cat. And this should admonish us to provide hogs with warm and comfortable lodging. Their sagacity in providing against cold in the night, when they have it in their power to make such provision, is quite wonderful. You see them looking about for the warmest spot: then they go to work, raking up the litter so as to break the wind off; and when they have done their best, they lie down. I had a sow that had some pigs running about with her in April last. There was a place open to her on each side of the barn. One faced the east and the other the west; and, I observed, that she sometimes took to one side and sometimes to the other. One evening her pigs had gone to bed on the east side. She was out eating till it began to grow dusk. I saw her go into her pigs, and was surprised to see her come out again; and therefore, looked a little to see what she was after. There was a high heap of dung in the front of the barn to the south. She walked up to the top of it, raised her nose, turned it very slowly, two or three times, from the north-east to the north-west, and back again, and at last, it settled at about south-east, for a little bit. She then came back, marched away very hastily to her pigs, roused them up in a great bustle, and away she trampled with them at her heels to the place on the west side of the barn. There was so little wind, that I could not tell which way it blew, till I took up some leaves, and tossed them in the air. I then found, that it came from the precise point which her nose had settled at. And thus was I convinced, that she had come out to ascertain which way the wind came, and, finding it likely to make her young ones cold in the night, she had gone and called them up, though it was nearly dark, and taken them off to a more comfortable birth. Was this an *instinctive*, or was it a *reasoning* proceeding? At any rate, let us not treat such animals as if they were stocks and stones.

309. POULTRY.—I merely mean to observe, as to poultry, that they must be kept away from turnips and cabbages, especially in the early part of the growth of these plants. When turnips are an inch or two high a good large flock of turkeys will destroy an acre in half a day, in four feet rows. Ducks and geese will do the same. Fowls will do great mischief. If these things cannot be kept out of the field, the crop must be abandoned, or the poultry killed. It is true, indeed, that it is only near the house that poultry plague

you much: but, it is equally true, that the best and richest land is precisely that which is near the house, and this, on every account, whether of produce or application, is the very land where you ought to have these crops.

CHAP. IX.

PRICES OF LAND, LABOUR, FOOD AND RAIMENT.

310. LAND is of various prices, of course. But, as I am, in this Chapter, addressing myself to *English Farmers,* I am not speaking of the price either of land in the *wildernesses,* or of land in the immediate vicinage of great cities. The wilderness price is two or three dollars an acre: the city price four or five hundred. The land at the same distance from New York that Chelsea is from London, is of higher price than the land at Chelsea. The surprizing growth of these cities, and the brilliant prospect before them, give value to every thing that is situated in or near them.

311. It is my intention, however, to speak only of *farming land.* This, too, is, of course, affected in its value by the circumstance of distance from market; but, the reader will make his own calculations as to this matter. A farm, then, on this Island, any where not nearer than thirty miles of, and not more distant than sixty miles from, New York, with a good farm-house, barn, stables, sheds, and styes; the land fenced

into fields with posts and rails, the wood-land being in the proportion of one to ten of the arable land, and there being on the farm a pretty good orchard; such a farm, if the land be in a good state, and of an average quality, is worth *sixty dollars an acre*, or *thirteen pounds sterling*; of course, a farm of a hundred acres would cost one thousand three hundred pounds. The rich lands on the *necks* and *bays*, where there are *meadows* and surprizingly productive orchards, and where there is *water carriage*, are worth, in some cases, three times this price. But, what I have said will be sufficient to enable the reader to form a pretty correct judgment on the subject. In New Jersey, in Pennsylvania, every where the price differs with the circumstances of water carriage, quality of land, and distance from market.

312. When I say a good farm-house, I mean a house *a great deal better* than the *general run* of farm-houses in England. More neatly finished on the inside. More in a *parlour* sort of style; though *round about* the house, things do not look so neat and tight as in England. Even in Pennsylvania, and amongst the Quakers too, there is a sort of out-of-doors slovenliness, which is never hardly seen in England. You see bits of wood, timber, boards, chips, lying about, here and there, and pigs and cattle

trampling about in a sort of confusion, which would make an English farmer fret himself to death; but which is here seen with great placidness. The out-buildings, except the barns, and except in the finest counties of Pennsylvania, are not so numerous, or so capacious, as in England, in proportion to the size of the farms. The reason is, that the *weather is so dry*. Cattle need not covering a twentieth part so much as in England, except hogs, who must be *warm* as well as dry. However, these share with the rest, and very little covering they get.

313. *Labour* is the great article of expence upon a farm; yet it is not nearly so great as in England, in proportion to the amount of the produce of a farm, especially if the poor-rates be, in both cases, included. However, speaking of the positive wages, a *good* farm-labourer has *twenty-five pounds sterling a year* and his board and lodging; and a *good* day-labourer has, upon an average, *a dollar a day*. A woman servant, in a farm-house, has from forty to fifty dollars a year, or eleven pounds sterling. These are the average of the wages throughout the country. But, then, mind, the farmer has nothing (for, really, it is not worth mentioning) to pay in *poor-rates*; which in England, must always be added to the wages

that a farmer pays; and, sometimes, they far exceed the wages.

314. It is, too, of importance to know, *what sort* of labourers these Americans are; for, though a labourer is a labourer, still there is some difference in them; and, these Americans are *the best that I ever saw*. They mow *four acres* of *oats, wheat, rye,* or *barley* in a day, and, with a cradle, lay it so smooth in the swarths, that it is tied up in sheaves with the greatest neatness and ease. They mow *two acres and a half of grass* in a day, and they do the work well. And the crops, upon an average, are all, except the wheat, *as heavy* as in England. The English farmer will want nothing more than these facts to convince him, that the labour, after all, is not so *very dear*.

315. The causes of these performances, so far beyond those in England, is first, the men are *tall* and well built; they are *bony* rather than *fleshy;* and they *live*, as to food, as well as man can live. And, secondly, they have been *educated* to do much in a day. The farmer here generally is at the *head* of his "*boys*," as they, in the kind language of the country, are called. Here is the best of examples. My old and beloved friend, Mr. JAMES PAUL, used, at the age of nearly *sixty* to go at *the head of his mowers*, though his fine farm was his own, and

though he might, in other respects, be called a rich man; and, I have heard, that Mr. ELIAS HICKS, the famous Quaker Preacher, who lives about nine miles from this spot, has this year, at *seventy* years of age, cradled down four acres of rye in a day. I wish some of the *preachers* of other descriptions, especially our fat parsons in England, would think a little of this, and would betake themselves to "work with their "hands the things which be good, that they "may have to give to him who needeth," and not go on any longer gormandizing and swilling upon the labour of those who need.

316. Besides the great quantity of work performed by the American labourer, his *skill*, the *versatility* of his talent, is a great thing. Every man can use an *ax*, a *saw*, and a *hammer*. Scarcely one who cannot do any job at rough carpentering, and mend a plough or a waggon. Very few indeed, who cannot kill and dress pigs and sheep, and many of them Oxen and Calves. Every farmer is a *neat* butcher; a butcher for *market;* and, of course, "the boys" must learn. This is a great convenience. It makes you so independent as to a main part of the means of housekeeping. All are *ploughmen.* In short, a good labourer here, can do *any thing* that is to be done upon a farm.

317. The operations necessary in miniature

cultivation they are very awkward at. The *gardens are ploughed* in general. An American labourer uses a *spade* in a very awkward manner. They *poke the earth about* as if they had no eyes; and toil and muck themselves half to death to dig as much ground in a day as a Surrey man would dig in about an hour of hard work. *Banking*, *hedging*, they know nothing about. They have no idea of the use of a *bill-hook*, which is so adroitly used in the coppices of Hampshire and Sussex. An *ax* is their tool, and with that tool, at *cutting down* trees or *cutting them up*, they will do *ten times* as much in a day as any other men that I ever saw. Set one of these men on upon a wood of timber trees, and his slaughter will astonish you. A neighbour of mine tells a story of an Irishman, who promised he could *do any thing*, and whom, therefore, to begin with, the employer sent into the wood to cut down a load of wood to burn. He staid a long while away with the team, and the farmer went to him fearing some accident had happened. "What are you about all this time?" said the farmer. The man was hacking away at a hickory tree, but had not got it half down; and that was all he had done. An American, black or white, would have had half a dozen trees cut down, cut up into lengths, put upon the carriage, and brought home, in the time.

318. So that our men, who come from England, must not expect, that, in these *common labours* of the country, they are to surpass, or even equal, these " *Yankees*," who, of all men that I ever saw, are the most *active* and the most *hardy*. They skip over a fence like a greyhound. They will catch you a pig in an open field by *racing* him down; and they are afraid of nothing. This was the sort of stuff that filled the *frigates* of DECATUR, HULL, and BRAINBRIDGE. No wonder that they triumphed when opposed to poor pressed creatures, worn out by length of service and ill-usage, and encouraged by no hope of fair-play. My LORD COCHRANE said, in his place in parliament, that it would be so; and so it was. Poor CASHMAN, that brave Irishman, with his dying breath, accused the government and the merchants of England of withholding from him his pittance of prize money! Ought not such a vile, robbing, murderous system to be destroyed?

319. Of the same active, hardy, and brave stuff, too, was composed the army of JACKSON, who drove the invaders into the Gulph of Mexico, and who would have driven into the same Gulph the army of Waterloo, and the heroic gentleman, too, who lent his hand to the murder of Marshal Ney. This is the stuff that stands between the rascals, called the Holy

Alliance, and the slavery of the whole civilized world. This is the stuff that gives us Englishmen an asylum; that gives us time to breathe; that enables us to deal our tyrants blows, which, without the existence of this stuff, they never would receive. This America, this scene of happiness under a free government, is the beam in the eye, the thorn in the side, the worm in the vitals, of every despot upon the face of the earth.

320. An American labourer is not regulated, as to time, by *clocks* and *watches*. The *sun*, who seldom hides his face, tells him when to begin in the morning and when to leave off at night. He has a dollar, a *whole dollar* for his work; but then it is the work of a *whole day*. Here is no dispute about *hours*. "*Hours* were "made for *slaves*," is an old saying; and, really, they seem here to act upon it as a practical maxim. This is a *great thing* in agricultural affairs. It prevents so many disputes. It removes so great a cause of disagreement. The American labourers, like the tavern-keepers, are never *servile*, but always *civil*. Neither *boobishness* nor *meanness* mark their character. They never *creep* and *fawn*, and are never *rude*. Employed about your house as day-labourers, they never come to interlope for victuals or drink. They have no idea of such a thing:

Their pride would restrain them if their plenty did not; and, thus would it be with all labourers, in all countries, were they left to enjoy the fair produce of their labour. Full pocket or empty pocket, these American labourers are always the *same men:* no saucy cunning in the one case, and no base crawling in the other. This, too, arises from the free institutions of government. A man has a voice *because he is a man,* and not because he is the *possessor of money.* And, shall I *never* see our English labourers in this happy state?

321. Let those English farmers, who love to see a poor wretched labourer stand trembling before them with his hat off, and who think no more of him than of a dog, remain where they are; or, go off, on the cavalry horses, to the devil at once, if they wish to avoid the tax-gatherer; for, they would, here, meet with so many mortifications, that they would, to a certainty, hang themselves in a month.

322. There are some, and even many, farmers, who *do not work themselves in the fields.* But, they all *attend* to the thing, and are all equally civil to their working people. They manage their affairs very judiciously. Little talking. Orders plainly given in few words, and in a decided tone. This is their only secret.

323. The *cattle* and *implements* used in hus-

bandry are cheaper than in England; that is to say, *lower priced.* The wear and tear not nearly half so much as upon a farm in England of the same size. The climate, the soil, the gentleness and docility of the horses and oxen, the lightness of the waggons and carts, the lightness and toughness of the *wood* of which husbandry implements are made, the simplicity of the harness, and, above all, the ingenuity and handiness of the workmen in *repairing*, and in *making shift;* all these make the implements a matter of very little note. Where horses are kept, the *shoing* of them is the most serious kind of expence.

324. The first business of a farmer is, here, and ought to be every where, to *live well:* to live in ease and plenty; to "*keep hospitality,*" as the old English saying was. To *save money* is a secondary consideration; but, any English farmer, who is a good farmer there, may, if he will bring his industry and care with him, and be *sure* to leave his pride and insolence (if he have any) along with his anxiety, behind him, live in ease and plenty here, and keep hospitality, and save a great parcel of money too. If he have the Jack-Daw taste for heaping little round things together in a hole, or chest, he may follow his taste. I have often thought of my good neighbour, JOHN GATER, who, if he were here, with his pretty clipped hedges,

his garden-looking fields, and his neat homesteads, would have visitors from far and near; and, while every one would admire and praise, no soul would envy him his possessions. Mr. GATER would soon have all these things. The hedges only want planting; and he would feel so comfortably to know that the Botley Parson could never again poke his nose into his sheepfold or his pig-stye. However, let me hope, rather, that the destruction of the Borough-tyranny, will soon make England a country, fit for an honest and industrious man to live in. Let me hope, that a relief from grinding taxation will soon relieve men of their fears of dying in poverty, and will, thereby, restore to England the "*hospitality*," for which she was once famed, but which now really exists no where but in America.

CHAP. X.

EXPENCES OF HOUSE-KEEPING.

325. It must be obvious, that these must be in proportion to the number in family, and to the style of living. Therefore, every one knowing how he stands in these two respects, the best thing for me to do is to give an account of the *prices* of house-rent, food, raiment, and servants; or, as they are called here, *helpers*.

326. In the great cities and towns house-rent is very high-priced; but, then, nobody but mad people live there except they have *business* there, and, then, they are paid back their rent in the *profits of that business.* This is so plain a matter, that no argument is necessary. It is unnecessary to speak about the expences of a *farm-house;* because, the farmer eats, and very frequently wears, his own produce. If these be high-priced, so is that part which he *sells.* Thus both ends meet with him.

327. I am, therefore, supposing the case of a man, who follows *no business,* and who lives upon what he has got. In England he cannot eat and drink and wear the interest of his money;

for the Boroughmongers have *pawned* half his income, and they will have it, or his blood. He wishes to escape from this alternative. He wishes to keep his blood, and enjoy his money too. He would come to America; but he does not know, whether prices here will not make up for the robbery of the Borough-villains; and he wishes to know, too, *what sort of society* he is going into. Of the latter I will speak in the *next chapter*.

328. The price of house-rent and fuel is, when at more than three miles from New York, as low as it is at the same distance from any great city or town in England. The price of wheaten bread is a third lower than it is in any part of England. The price of *beef, mutton, lamb, veal, small pork, hog-meat, poultry,* is *one half the London price;* the first as good, the two next very nearly as good, and all the rest far, very far, better than in London. The sheep and lambs that I now kill for my house are as fat as any that I ever saw in all my life; and they have been running in *wild ground,* wholly uncultivated for many years, all the summer. A lamb, killed the week before last, weighing in the whole, *thirty-eight pounds,* had five *pounds of loose fat* and *three pounds and ten ounces of suet.* We cut a pound of solid fat from each breast; and, after that it was too

fat to be pleasant to eat. My flock being small, forty, or thereabouts, of some neighbours joined them; and they have all got fat together. I have missed the interlopers lately: I suppose the "Yorkers" have eaten them up by this time. What they have fattened on except *brambles* and *cedars*, I am sure I do not know. If any Englishman should be afraid that he will find no roast-beef here, it may be sufficient to tell him, that an ox was killed, last winter, at Philadelphia, the quarters of which weighed *two thousand, two hundred, and some odd pounds*, and he was sold TO THE BUTCHER for *one thousand three hundred dollars*. This is proof enough of the spirit of enterprize, and of the disposition in the public to encourage it. I believe this to have been the *fattest* ox that ever was killed in the world. Three times as much money, or, perhaps, ten times as much, might have been made, if the ox had been *shown for money*. But, this the owner *would not permit*; and he sold the ox in that condition. I need hardly say that the owner was a Quaker. New Jersey had the honour of producing this ox, and the owner's name was JOB TYLER.

329. That there must be good *bread* in America is pretty evident from the well known fact, that hundreds of thousands of barrels of flour are, most years sent to England, finer than any

that England can produce. And, having now provided the two principal articles, I will suppose, as a matter of course, that a gentleman will have *a garden*, an *orchard*, and a *cow* or two; but, if he should be able (no easy matter) to find a genteel country-house without these conveniences, he may buy *butter*, cheaper, and, upon an average, better than in England. The garden stuff, if he send to New York for it, he must buy pretty dear; and, faith, he *ought* to buy it dear, if he will not have some planted and preserved.

330. *Cheese*, of the North River produce, I have bought as good of Mr. STICKLER of New York as I ever tasted in all my life; and, indeed, no better cheese need be wished for than what is now made in this country. The average price is about *seven pence a pound* (English money), which is much lower than even *middling* cheese is in England. Perhaps, *generally speaking*, the cheese here is not so good as the better kinds in England; but, there is none here so poor as the poorest in England. Indeed the people *would not eat it*, which is the best security against its being made. Mind, I state distinctly, that as good cheese as I ever tasted, if not the best, was of American produce. I know the article well. Bread and cheese *dinners* have been the dinners of a good fourth of

my life. I know the Cheshire, Gloucester, Wiltshire, Stilton, and the Parmasan; and I never tasted better than American cheese, bought of Mr. STICKLER, in Broad Street, New York. And this cheese Mr. STICKLER informs me is nothing uncommon in the county of Cheshire in Massachusetts; he knows at least a hundred persons himself that make it equally good. And, indeed, why should it not be thus in a country where the pasture is so rich; where the *sun* warms every thing into sweetness; where the cattle eat the grass close *under the shade of the thickest trees;* which we know well they will not do in England. Take any fruit which has grown in the shade in England, and you will find that it has not half the sweetness in it, that there is in fruit of the same bulk, grown in the sun. But, here the sun sends his heat down through all the boughs and leaves. The *manufacturing* of cheese is not yet *generally* brought, in this country, to the English perfection; but, here are all the materials, and the rest will soon follow.

331. *Groceries*, as they are called, are, upon an average, at far less than *half* the English price. Tea, sugar, coffee, spices, chocolate, cocoa, salt, sweet oil; all free of the Boroughmongers' *taxes* and their *pawn*, are so cheap as to be within the reach of every one. Chocolate,

which is a *treat* to the *rich*, in England, is here used even by *the negroes*. Sweet oil, raisins, currants; all the things from the Levant, are at a *fourth* or *fifth* of the English price. The English people, who pay enormously to keep possession of the East and West Indies, purchase the produce even of the English possessions at a price double of that which the Americans give *for that very produce!* What a hellish oppression must that people live under! Candles and soap (quality for quality) are half the English price. Wax candles (beautiful) are at a *third* of the English price. It is no very great piece of extravagance to burn wax candles *constantly* here, and it is frequently done by genteel people, who do not make their own candles.

332. *Fish* I have not mentioned, because fish is not *every where* to be had in abundance. But, any where near the coast it is; and, it is so cheap, that one wonders how it can be brought to market for the money. Fine Black-Rock, as good, at least, as codfish, I have seen sold, and in cold weather too, at an *English farthing a pound*. They now bring us fine fish round the country to our doors, at an English three pence a pound. I believe they count *fifty* or *sixty sorts* of fish in New York market, as the average. Oysters, other shell-fish, called

clams. In short, the variety and abundance are such that I cannot describe them.

333. An idea of the state of *plenty* may be formed from these facts: nobody but the free negroes who have families ever think of eating a *sheep's head and pluck.* It is seldom that *oxen's heads* are used at home, or *sold,* and never in the country. In the course of the year hundreds of *calves' heads,* large bits and *whole joints* of meat, are left on the shambles, at New York, for any body *to take away* that will. They generally fall to the share of the *street hogs,* a thousand or two of which are constantly *fatting* in New York on the meat and fish flung out of the houses. I shall be told, that it is only in *hot weather,* that the shambles are left thus garnished. Very true; but, are the shambles of *any other country* thus garnished in *hot weather?* Oh! no! If it were not for the superabundance, all the food would be sold at *some* price or other.

334. After bread, flesh, fish, fowl, butter, cheese and groceries, comes *fruit.* Apples, pears, cherries, peaches at a *tenth* part of the English price. The other day I met a man going to market with a waggon load of *winter pears.* He had high boards on the sides of the waggon, and his waggon held about 40 or 50 bushels. I have bought very good apples this

year for *four pence half penny* (English) a bushel, to boil for little pigs. Besides these, strawberries grow wild in abundance; but no one will take the trouble to get them. Huckle-berries in the woods in great abundance, chesnuts all over the country. Four pence half-penny (English) a quart for these latter. Cranberries, the finest fruit for tarts that ever grew, are bought for about a dollar a bushel, and they will keep, flung down in the corner of a room, for five months in the year. As a sauce to venison or mutton, they are as good as currant jelly. Pine apples in abundance, for several months in the year, at an average of an English shilling each. Melons at an average of an English eight pence. In short, what is there not in the way of fruit? All excellent of their kinds and all for a mere trifle, compared to what they cost in England.

335. I am afraid to speak of *drink*, lest I should be supposed to countenance the *common use* of it. But, protesting most decidedly against this conclusion, I proceed to inform those, who are not content with the *cow* for vintner and brewer, that all the materials for making people drunk, or muddle headed, are much cheaper here than in England. Beer, *good ale*, I mean, a great deal better than the common public-house beer in England; in short, good, strong, clear ale, is, at New York, eight dollars a bar-

rel; that is, about *fourteen English pence a gallon.* Brew yourself, in the country, and it is about *seven English pence a gallon;* that is to say, *less than two pence a quart.* No Borough-mongers' tax on malt, hops, or beer! Portugal wine is about *half* the price that it is in England. French wine a *sixth part* of the English price. Brandy and Rum about the same in proportion; and the common spirits of the country are about three shillings and sixpence (English) *a gallon.* Come on, then, if you love toping; for here you may drink yourselves blind at the price of sixpence.

336. WEARING APPAREL comes chiefly from England, and all the *materials* of dress are as cheap as they are there; for, though there is a duty laid on the importation, the absence of taxes, and the cheap food and drink, enable the retailer to sell as low here as there. Shoes are cheaper than in England; for, though shoe-makers are well paid for their labour, there is no Borough-villain to *tax the leather.* All the *India* and *French* goods are at half the English price. Here no ruffian can seize you by the throat and tear off your suspected handkerchief. Here SIGNOR WAITHMAN, or any body in that line, might have sold French gloves and shawls without being tempted to quit the field of politics as a compromise with the government; and

without any breach of covenants, after being suffered to escape with only a gentle squeeze.

337. *Household Furniture*, all cheaper than in England. *Mahogany* timber a third part of the English price. The distance shorter to bring it, and the tax next to nothing on importation. The *woods* here, the pine, the ash, the white-oak, the walnut, the tulip-tree, and many others, all excellent. The workman paid high wages, but *no tax*. No Borough-villains to share in the amount of the price.

338. Horses, carriages, harness, all as good, as gay, and cheaper than in England. I hardly ever saw a *rip* in this country. The hackney coach horses and the coaches themselves, at New York, bear no resemblance to things of the same name in London. The former are all good, sound, clean, and handsome. What the latter are I need describe in no other way than to say, that the coaches seem fit for nothing but the fire and the horses for the dogs.

339. *Domestic servants!* This is a weighty article: not in the *cost*, however, so much as in the plague. A *good man servant* is worth *thirty pounds sterling* a year; and a *good woman servant, twenty pounds sterling a year*. But, this is not all; for, in the first place, they will hire only *by the month*. This is what they, in fact, do in England; for, there they can quit

at a *month's warning*. The man will not wear *a livery*, any more than he will wear a halter round his neck. This is no great matter; for, as your neighbours' men are of the same taste, you expose yourself to no humiliation on this score. Neither men nor women will allow you to call them *servants*, and they will take especial care not to call themselves by that name. This seems something very capricious, at the least; and, as people in such situations of life, really *are* servants, according to even the sense which MOSES gives to the word, when he forbids the working of the *man servant* and the *maid servant*, the objection, the rooted aversion, to the name, seems to bespeak a mixture of *false pride* and of *insolence*, neither of which belong to the American character, even in the lowest walks of life. I will, therefore, explain the *cause* of this dislike to the name of servant. When this country was first settled, there were no people that *laboured for other people;* but, as man is always trying to throw the working part off his own shoulders, as we see by the conduct of *priests* in all ages, *negroes* were soon introduced. *Englishmen*, who had fled *from tyranny* at home, were naturally shy of calling other men their *slaves;* and, therefore, "*for more grace*," as Master Matthew says in the play, they called their slaves *servants*. But, though I doubt not that this device was quite efficient in quieting their own consciences, it gave rise to the notion, that *slave* and *servant* meant one and the same thing, a conclusion perfectly natural and directly deducible from the premises. Hence every *free* man and woman have rejected with just disdain the appellation of *servant*. One would think, however, that they might be reconciled to it by the conduct of some of their superiors in life, who, without the smallest apparent reluctance, call themselves "*Public Servants*," in imitation, I suppose, of English Ministers, and his Holiness, the Pope, who, in the excess of his humility, calls himself, "*the* "*Servant of the Servants of the Lord.*" But, perhaps, the American Domestics have observed, that "*Public Servant*" really means *master*. Be the cause what it may, however, they continue most obstinately to scout the name of servant; and, though they still keep a civil tongue in their head, there is not one of them that will not resent the affront with more bitterness than any other that you can offer. The man, therefore, who would deliberately offer such an affront must be a fool. But, there is an inconvenience far greater than this. People in general are so comfortably situated, that very few, and then only of those who are pushed hard, will become domestics to any body. So that, generally speaking, Domestics of both sexes are far from good. They are *honest;* but they are not *obedient.* They are careless. Wanting frequently in the greater part of those qualities, which make their services conducive to the neatness of houses and comfort of families. What a difference would it make in this country, if it could be supplied with nice, clean, dutiful English maid servants! As to the *men,* it does not much signify; but, for the want of the maids, nothing but the absence of grinding taxation can compensate. As to *bringing them with you,* it is as wild a project as it would be to try to carry the sunbeams to England. They will begin to change before the ship gets on soundings; and, before they have been here a month, you must turn them out of doors, or they will you. If, by any chance, you *find them here,* it may do; but bring them out and keep them you cannot. The best way is to put on your philosophy; never to look at this evil without, at the same time, looking at the many good things that you find here. Make the best selection you can. Give *good wages,* not too much work, and resolve, at all events, to treat them with *civility.*

340. However, what is this plague, compared with that of the *tax gatherer?* What is this plague compared with the constant sight of

beggars and paupers, and the constant dread of becoming a pauper or beggar yourself? If your commands are not obeyed with such alacrity as in England, you have, at any rate, nobody to *command you.* You are not ordered to "*stand and deliver*" twenty or thirty times in the year by the insolent agent of Boroughmongers. No one comes to forbid you to open or shut up a window. No insolent set of Commissioners send their order for you to dance attendance on them, to *shew cause* why they should not *double-tax you;* and, when you have shown cause, even on your oath, make you pay the tax, laugh in your face, and leave you *an appeal* from themselves to another set, deriving their authority from the same source, and having a similar interest in oppressing you, and thus laying your property prostrate beneath the hoof of an insolent and remorseless tyranny. Free, wholly free, from this tantalizing, this grinding, this odious curse, what need you care about the petty plagues of Domestic Servants?

341. However, as there are some men and some women, who can never be at heart's ease, unless they have the power of domineering over somebody or other, and who will rather be slaves themselves than not have it in their power to treat others as slaves, it becomes a man of

fortune, proposing to emigrate to America, to consider soberly, whether he, or his wife, be of this taste; and, if the result of his consideration be in the affirmative, his best way will be to continue to live under the Boroughmongers, or, which I would rather recommend, hang himself at once.

CHAP. XI.

MANNERS, CUSTOMS, AND CHARACTER OF THE PEOPLE.

342. ALL these are, generally speaking, the same as those of the people of England. The French call this people *Les Anglo-Americains*; and, indeed, what are they else? Of the manners and customs somewhat peculiar to America I have said so much, here and there, in former Chapters, that I can hardly say any thing new here upon these matters. But, as *society* is naturally a great thing with a gentleman, who thinks of coming hither with his wife and children, I will endeavour to describe the society that he will find here. To give *general* descriptions is not so satisfactory as it is to deal a little in particular instances; to tell of what one has seen and experienced. This is what I shall do; and, in this Chapter I wish to be regarded as addressing myself to a most worthy and public-spirited gentleman of moderate fortune, *in Lancashire*, who, with a large family, now balances whether he shall come, or stay.

343. Now, then, my dear Sir, this people contains very few persons very much raised in men's estimation, above the general mass; for, though there are some men of immense *fortunes*, their wealth does very little indeed in the way of purchasing even the outward signs of respect; and, as to *adulation*, it is not to be purchased with love or money. Men, be they what they may, are generally called by their *two names*, without any thing prefixed or added. I am one of the greatest men in this country at present; for people in general call me "*Cobbett*," though the Quakers provokingly persevere in putting the *William* before it, and my old friends in Pennsylvania, use even the word *Billy*, which, in the very sound of the letters, is an antidote to every thing like thirst for distinction.

344. Fielding, in one of his romances, observes, that there are but few cases, in which a husband can be justified in availing himself of the right which the law gives him to bestow manual chastisement upon his wife, and that one of these, he thinks, is, when any pretensions to *superiority of blood* make their appearance in her language and conduct. They have a better cure for this malady here; namely; silent, but, *ineffable contempt*.

345. It is supposed, in England, that this equality of estimation must beget a general

coarseness and rudeness of behaviour. Never was there a greater mistake. No man likes to be treated with disrespect; and, when he finds that he can obtain respect only by treating others with respect, he will use that only means. When he finds that neither haughtiness nor wealth will bring him a civil word, he becomes civil himself; and, I repeat it again and again, this is a country of *universal civility*.

346. The causes of *hypocrisy* are the fear of loss and the hope of gain. Men crawl to those, whom, in their hearts, they despise, because they fear the effects of their ill-will and hope to gain by their good-will. The circumstances of all ranks are so easy here, that there is no cause for hypocrisy; and the thing is not of so fascinating a nature, that men should love it for its own sake.

347. The boasting of wealth, and the endeavouring to disguise poverty, these two acts, so painful to contemplate, are almost total strangers in this country; for, no man can gain adulation or respect by his wealth, and no man dreads the effects of poverty, because no man sees any dreadful effects arising from poverty.

348. That *anxious eagerness to get on*, which is seldom unaccompanied with some degree of *envy* of more successful neighbours, and which has its foundation first in *a dread of future want*,

and next in a *desire to obtain distinction by means of wealth;* this anxious eagerness, so unamiable in itself, and so unpleasant an inmate of the breast, so great a sourer of the temper, is a stranger to America, where accidents and losses, which would drive an Englishman half mad, produce but very little agitation.

349. From the absence of so many causes of uneasiness, of envy, of jealousy, of rivalship, and of mutual dislike, *society*, that is to say, the intercourse between man and man, and family and family, becomes easy and pleasant; while the universal plenty is the cause of universal hospitality. I know, and have ever known, but little of the people in the cities and towns in America; but, the difference between them and the people in the country can only be such as is found in all other countries. As to the manner of living in the country, I was, the other day, at a gentleman's house, and I asked the lady for *her bill of fare for the year.* I saw *fourteen* fat hogs, weighing about *twenty score a piece*, which were to come *into the house* the next Monday; for here they slaughter them *all in one day.* This led me to ask, "Why, in "God's name, what do you eat in a year?" The Bill of fare was this, for this present year: about *this same quantity of hog-meat; four beeves;* and *forty-six fat sheep!* Besides the

sucking pigs (of which we had then one on the table), besides *lambs*, and besides the produce of *seventy hen fowls*, not to mention good parcels of *geese*, *ducks* and *turkeys*, but, not to forget a garden of three quarters of an acre and *the butter of ten cows*, not one ounce of which is ever *sold!* What do you think of that? Why, you will say, this must be some *great overgrown farmer*, that has swallowed up half the country; or some nabob sort of merchant. Not at all. He has only *one hundred and fifty four acres of land*, (all he consumes is of the produce of this land), and he lives in the same house that his English-born grandfather lived in.

350. When the hogs are killed, the house is full of work. The sides are salted down as pork. The hams are smoked. The lean meats are made into sausages, of which, in this family, they make about *two hundred weight*. These latter, with broiled fish, eggs, dried beef, dried mutton, slices of ham, tongue, bread, butter, cheese, short cakes, buckwheat cakes, sweet meats of various sorts, and many other things, make up the *breakfast* fare of the year, and, a dish of *beef steakes* is frequently added.

351. When one sees this sort of living, with the houses *full of good beds*, ready for the guests as well as the family to sleep in, we can-

not help perceiving, that this is that "*English* "*Hospitality*," of which we have *read* so much; but, which Boroughmongers' taxes and pawns have long since driven out of England. This American way of life puts one in mind of FORTESCUE's fine description of the happy state of the English, produced by their *good laws*, which kept every man's property sacred, even from the grasp of the king. "Every in-"habitant is at his Liberty fully to use and en-"joy whatever his Farm produceth, the Fruits "of the Earth, the Increase of his Flock, and "the like: All the Improvements he makes, "whether by his own proper Industry, or of "those he retains in his Service, are his own to "use and enjoy without the Lett, Interruption, "or Denial of any: If he be in any wise in-"jured, or oppressed, he shall have his *Amends* "and Satisfaction against the party offending: "Hence it is, that the Inhabitants are Rich in "Gold, Silver, and in all the Necessaries and "Conveniences of Life. They drink no Water, "unless at certain Times, upon a Religious "Score, and by Way of doing Penance. They "are fed, in great Abundance, with all sorts of "Flesh and Fish, of which they have Plenty "every where; they are cloathed throughout "in good Woollens; their Bedding and other "Furniture in their Houses are of Wool, and

" that in great Store: They are also well pro-
" vided with all other Sorts of Household
" Goods, and necessary Implements for Hus-
" bandry: Every one, according to his Rank,
" hath all Things *which conduce to make Life*
" *easy and happy*. They are not sued at Law
" but before the Ordinary Judges, where they
" are treated with Mercy and Justice, accord-
" ing to the Laws of the Land; neither are
" they impleaded in Point of Property, or ar-
" raigned for any Capital Crime, how heinous
" soever, but before the King's Judges, and ac-
" cording to the Laws of the Land. These are
" the Advantages consequent from that *Politi-*
" *cal Mixt Government* which obtains in *Eng-*
" *land* ——"

352. This passage, which was first pointed out to me by SIR FRANCIS BURDETT, describes the state of England four hundred years ago; and this, with the *polish* of modern times added, is now the state of the Americans. Their forefathers brought the "English Hospitality" with them; for, when they left the country, the infernal *Boroughmonger Funding system* had not begun. The STUARTS were *religious* and *prerogative* tyrants; but they were not, like their successors, the Boroughmongers, taxing, plundering tyrants. Their quarrels with their subjects were about mere *words:* with the

Boroughmongers it is a question of purses and strong-boxes, of goods and chattels, lands and tenements. "*Confiscation*" is their word; and you must submit, be hanged, or flee. They take away men's property at their pleasure, *without any appeal to any tribunal.* They appoint *Commissioners* to seize what they choose. There is, in fact, *no law* of property left. The Bishop-begotten and hell-born system of Funding has stripped England of every vestige of what was her ancient character. Her hospitality along with her freedom have crossed the Atlantic; and here they are to shame our ruffian tyrants, if they were sensible of shame, and to give shelter to those who may be disposed to deal them distant blows.

353. It is not with a little bit of dry toast, so neatly put in a rack; a bit of butter so round and small; a little milk pot so pretty and so empty; an egg *for you*, the host and hostess *not liking eggs.* It is not with looks that seem to say, "don't eat too much, for the taxgatherer "is coming." It is not thus that you are received in America. You are not much *asked*, not much *pressed*, to eat and drink; but, such an abundance is spread before you, and so hearty and so cordial is your reception, that you instantly lose all restraint, and are tempted

to feast whether you be hungry or not. And, though the *manner* and *style* are widely different in different houses, the *abundance* every where prevails. This is the strength of the government: a happy people: and no government ought to have any other strength.

354. But, you may say, perhaps, that plenty, however great, is not *all* that is wanted. Very true: for the *mind* is of more account than the carcass. But, here is mind too. These repasts, amongst people of any figure, come forth under the superintendance of industrious and accomplished house-wifes, or their daughters, who all *read a great deal,* and in whom that gentle treatment from parents and husbands, which arises from an absence of racking anxiety, has created an habitual, and even an hereditary *good humour*. These ladies can converse with you upon almost any subject, and the ease and gracefulness of their behaviour are surpassed by those of none of even our best-tempered English women. They fade at an earlier age than in England; but, till then, they are as beautiful as the women in *Cornwall,* which contains, to my thinking, the prettiest women in our country. However, young or old, blooming or fading, well or ill, rich or poor, they still preserve their *good humour*.

"But, since, alas! frail beauty must decay,
"Curl'd, or uncurl'd, since locks will turn to grey;
"Since painted, or not painted, all shall fade,
"And she who scorns a man must die a maid;
"What, then, remains, but well our pow'r to use,
"And keep *good humour* still, whate'er we lose?
"And, trust me, Dear, good-humour can prevail,
"When flights and fits, and screams and scolding fail."

355. This beautiful passage, from the most beautiful of poets, which ought to be fastened in large print upon every lady's dressing table, the American women, of all ranks, seem to have by heart. Even amongst the very lowest of the people, you seldom hear of that torment, which the old proverb makes the twin of a smoky house.

356. There are very few really *ignorant* men in America of native growth. Every farmer is more or less of *a reader*. There is no *brogue*, no *provincial dialect*. No class like that which the French call *peasantry*, and which degrading appellation the miscreant spawn of the Funds have, of late years, applied to the whole mass of the most useful of the people in England, those who do the work and fight the battles. And, as to the men, who would naturally form *your* acquaintances, they, I know from experience, are as kind, frank, and sensible men as are, on the general run, to be found in England, even with the power of selection. They are all well-

informed; modest without shyness; always free to communicate what they know, and never ashamed to acknowledge that they have yet to learn. You never hear them *boast* of their possessions, and you never hear them *complaining* of their wants. They have all been *readers* from their youth up; and there are few subjects upon which they cannot converse with you, whether of a political or scientific nature. At any rate, they always *hear* with patience. I do not know that I ever heard a native American interrupt another man while he was speaking. Their *sedateness* and *coolness,* the *deliberate* manner in which they say and do every thing, and the *slowness* and *reserve* with which they express their assent; these are very wrongly estimated, when they are taken for marks of *a want of feeling*. It must be a tale of woe indeed, that will bring a tear from an American's eye; but any trumped up story will send his hand to his pocket, as the ambassadors from the beggars of France, Italy and Germany can fully testify.

357. However, you will not, for a long while, know what to do for want of the *quick responses* of the English tongue, and the *decided* tone of the English expression. The *loud voice;* the *hard squeeze* by the hand; the *instant assent or dissent;* the *clamorous joy;* the *bitter wailing;*

the *ardent friendship*; the *deadly enmity*; the *love that makes people kill themselves*; the *hatred that makes them kill others.* All these belong to the characters of Englishmen, in whose minds and hearts every feeling exists in the *extreme.* To decide the question, which character is, upon the whole, *best*, the American or the English, we must appeal to some *third party.* But, it is no matter: we cannot change our natures. For my part, who can, in nothing, think or act by halves, I must belie my very nature, if I said that I did not like the character of my own countrymen best. We all like our own parents and children better than other people's parents and children; not because they *are* better, but because they are *ours;* because they belong to us and we to them, and because we must *resemble* each other. There are some Americans that I like full as well as I do any man in England; but, if, nation against nation, I put the question home to my heart, it instantly decides in favour of my countrymen.

358. You must not be offended if you find people here take but little interest in the concerns of England. Why should they? BOLTON F———R cannot hire spies to entrap them. As matter of curiosity, they may contemplate such works as those of FLETCHER; but, they cannot *feel* much upon the subject; and

they are not insincere enough to express much.

359. There is one thing in the Americans, which, though its proper place was further back, I have reserved, or rather *kept back*, to the last moment. It has presented itself several times; but I have turned from the thought, as men do from thinking of any mortal disease that is at work in their frame. It is not covetousness; it is not niggardliness; it is not insincerity; it is not enviousness; it is not cowardice, above all things: it is DRINKING. Aye, and that too, amongst but too many men, who, one would think, would loath it. You can go into hardly any man's house, without being asked to drink wine, or spirits, even *in the morning*. They are quick at meals, are little eaters, seem to care little about what they eat, and never talk about it. This, which arises out of the universal abundance of good and even fine eatables, is very amiable. You are here disgusted with none of those *eaters by reputation* that are found, especially amongst *the Parsons*, in England: fellows that *unbutton* at it. Nor do the Americans *sit and tope much after dinner*, and talk on till they get into nonsense and *smut*, which last is a sure mark of a silly and, pretty generally, even of a base mind. But, they *tipple;* and the infernal spirits they

tipple too! The scenes that I witnessed at Harrisburgh I shall never forget. I almost wished (God forgive me!) that there were Boroughmongers here to *tax* these drinkers: they would soon reduce them to a moderate dose. Any nation that feels itself uneasy with its fulness of good things, has only to resort to an application of Boroughmongers. These are by no means nice feeders or of contracted throat: they will suck down any thing from the poor man's pot of beer to the rich man's lands and tenements.

360. The Americans preserve their gravity and quietness and good-humour even in their drink; and so much the worse. It were far better for them to be as noisy and quarrelsome as the English drunkards; for then the odiousness of the vice would be more visible, and the vice itself might become less frequent. Few vices want an *apology*, and drinking has not only its apologies but its *praises;* for, besides the appellation of "*generous wine*," and the numerous songs, some in very elegant and witty language, from the pens of debauched men of talents, drinking is said to be necessary, in certain cases at least, *to raise the spirits*, and *to keep out cold.* Never was any thing more false. Whatever intoxicates must *enfeeble* in the end,

and whatever enfeebles must *chill.* It is very well known, in the Northern countries, that, if the cold be such as to produce danger of *frost-biting*, you must take care *not to drink strong liquors.*

361. To see this beastly vice in *young men* is shocking. At one of the taverns at Harrisburgh there were several as fine young men as I ever saw. Well-dressed, well educated, polite, and every thing but *sober.* What a squalid, drooping, sickly set they looked *in the morning!*

362. Even little boys at, or under, *twelve* years of age, go into *stores,* and tip off their *drams!* I never struck a child, in anger, in my life, that I recollect; but, if I were so unfortunate as to have a son to do this, he having had an example to the contrary in me, I would, if all other means of reclaiming him failed, whip him like a dog, or, which would be better, make him an out-cast from my family.

363. However, I must not be understood as meaning, that this tippling is *universal* amongst gentlemen; and, God be thanked, the *women* of any figure in life do by no means give into the practice; but, abhor it as much as well-bred women in England, who, in general, no more think of drinking strong liquors, than they do of drinking poison.

364. I shall be told, that men in the *harvest field* must have *something* to drink. To be sure, where perspiration almost instantly carries off the drink, the latter does not remain so long to burn the liver, or whatever else it does burn. But, I much question the utility even here; and I think, that, in the long run, a water-drinker would beat a spirit drinker at any thing, provided both had plenty of good food. And, besides, *beer*, which does not *burn*, at any rate, is within every one's reach in America, if he will but take the trouble to brew it.

365. A man, at Botley, whom I was very severely reproaching for getting drunk and lying in the road, whose name was JAMES ISAACS, and who was, by the by, one of the hardest workers I ever knew, said, in answer, "Why, "now, Sir, NOAH and LOT were two very good "men, you know, and yet they loved *a drop of* "*drink*." "Yes, you drunken fool," replied I, "but you do not read that *Isaac* ever got "drunk and rolled about the road." I could not help thinking, however, that the BIBLE SOCIETIES, with the wise Emperor Alexander and the Holy Alliance at their head, might as well (to say nothing about the *cant* of the thing) leave the Bible to work its own way. I had seen ISAACS dead drunk, lying stretched out,

by my front gate, against the public highway; and, if he had followed the example of NOAH, he would not have endeavoured to excuse himself in the modest manner that he did, but would have affixed an *everlasting curse on me and my children to all generations.*

366. The soldiers, in the regiment that I belonged to, many of whom served in the American war, had a saying, that the *Quakers* used the word *tired* in place of the word *drunk*. Whether any of them do ever get *tired* themselves, I know not; but, at any rate they most resolutely set their faces against the common use of spirits. They forbid their members to retail them; and, in case of disobedience, they *disown* them.

367. However, there is no remedy but the introduction of *beer*, and, I am very happy to know, that beer is, every day, becoming more and more fashionable. At Bristol in Pennsylvania, I was pleased to see excellent beer in clean and nice pewter pots. Beer does not kill. It does not eat out the vitals and take the colour from the cheek. It will make men "*tired*," indeed, by midnight; but it does not make them half dead in the morning. We call wine the *juice of the grape*, and such it is with a proportion of *ardent spirits*, equal, in Portugal

wine, to a *fifth* of the wine; and, therefore, when a man has taken down a bottle of Port or of Madeira, he has nearly *half a pint* of ardent spirits in him. And yet how many foolish mothers give their children Port wine to *strengthen* them! I never like your *wine-physicians*, though they are great favourites with but too many patients. BONIFACE, in the *Beaux Stratagem*, says that he has eaten his ale, drunk his ale, worked upon his ale, and slept upon his ale, for forty years, and that he has grown fatter and fatter; but, that his wife (God rest her soul!) would not take it *pure:* she would adulterate it with brandy; till, at last, finding that the poor woman was never well, he put a tub of her favourite by her bedside, which, in a short time, brought her "a *happy release*" from this "state of probation," and carried her off into the "*the world of spirits.*" Whether Boniface meant this as a *pun*, I do not know; for, really, if I am to judge from the *practice* of many of the vagrant fanatics, I must believe, that, when they rave about the *spirit's entering them*, they mean that which goes out of a glass down their throat. Priests may make what they will of their devil; they may make him a reptile with a forked tongue, or a beast with a cloven hoof; they may, like Milton, dress him out with seraphic wings; or like Saint Francis, they may give him horns and tail: but, I say that the devil, who is the strongest tempter, and who produces the most mischief in the world, approaches us in the shape of *liquid*, not melted brimstone, but wine, gin, brandy, rum, and whiskey. One comfort is, however, that *this* devil, of whose existence we can have no doubt, who is visible and even tangible, we can, if we will, without the aid of priests, or, rather, in spite of them, easily and safely set at defiance. There are many wrong things which men do against the general and natural bent of their minds. Fraud, theft, and even murder, are frequently, and most frequently, the offspring of *want*. In these cases, it is a choice of evils; *crime* or *hunger*. But, drinking to excess is a man's own act; an evil deliberately sought after; an act of violence committed against reason and against nature; and that, too, without the smallest temptation, except from that vicious appetite, which he himself has voluntarily created.

368. You, my dear Sir, stand in need of no such lectures as this, and the same is, I hope, the case with the far greater part of my readers; but, if it tend, in the smallest degree, to check the fearful growth of this tree of

unmixed evil; if it should make the bottle less cherished even in one small circle; nay, if it keep but one young man in the world in the paths of sobriety, how could my time have been better bestowed?

CHAP. XII.

RURAL SPORTS.

369. THERE are persons, who question *the right* of man to pursue and destroy the wild animals, which are called *game*. Such persons, however, claim the right of killing *foxes* and *hawks;* yet, these have as much right to live and to follow their food as *pheasants* and *partridges* have. This, therefore, in such persons, is *nonsense.*

370. Others, in their mitigated hostility to the sports of the field, say, that it is *wanton* cruelty to shoot or hunt; and that we *kill* animals from the farm-yard only because their flesh *is necessary to our own existence.* PROVE THAT. No: you cannot. If you could, it is but the "*tyrant*'s plea;" but you cannot: for we know that men can, and do, live without animal food, and, if their labour be not of an exhausting kind, live well too, and longer than those who eat it. It comes to this, then, that we kill hogs and oxen because we *choose* to kill them; and, we kill game for precisely the same reason.

371. A third class of objectors, seeing the

weak position of the two former, and still resolved to eat flesh, take their stand upon this ground: that sportsmen send some game off *wounded* and leave them in *a state of suffering.* These gentlemen forget the operations performed upon calves, pigs, lambs and sometimes on poultry. Sir ISAAC COFFIN prides himself upon teaching the English ladies how to make *turkey-capons!* Only think of the separation of calves, pigs, and lambs, at an early age, from their mothers! Go, you sentimental eaters of veal, sucking pig and lamb, and hear the mournful lowings, whinings, and bleatings; observe the anxious listen, the wistful look, and the dropping tear, of the disconsolate dams; and, then, while you have the carcasses of their young ones under your teeth, cry out, as soon as you can empty your mouths a little, against the *cruelty* of hunting and shooting. Get up from dinner (but take care to stuff well first), and go and drown the puppies of the bitch, and the kittens of the cat, lest they should share a little in what their mothers have guarded with so much fidelity; and, as good stuffing may tend to make you restless in the night, order the geese to be picked alive, that, however your consciences may feel, your bed, at least, may be easy and soft. Witness all this with your own eyes; and then go weeping to bed, at the possibility of a hare having been terribly frightened without being killed, or of a bird having been left in a thicket with a shot in its body or a fracture in its wing. But, before you go up stairs, give your servant orders to be early at market for fish, fresh out of the water; that they may be *scaled*, or *skinned alive!* A truce with you, then, sentimental eaters of flesh: and here I propose the terms of a lasting compromise with you. We must, on each side, yield something: we sportsmen will content ourselves with merely *seeing the hares skip and the birds fly;* and you shall be content with the flesh and fish that come from cases of *natural death*, of which, I am sure, your compassionate disposition will not refuse us a trifling allowance.

372. Nor have even the *Pythagoreans* a much better battery against us. Sir RICHARD PHILLIPS, who once rang a peal in my ears against shooting and hunting, does, indeed, eat neither *flesh*, *fish*, nor *fowl*. His abstinence surpasses that of a Carmelite, while his bulk would not disgrace a Benedictine Monk, or a Protestant Dean. But, he forgets, that his *shoes* and *breeches* and *gloves* are made of the skins of animals: he forgets that he *writes* (and very eloquently too) with what has been cruelly taken from a fowl; and that, in order to cover

the *books* which he has had made and sold, hundreds of flocks and scores of droves must have perished: nay, that, to get him his *beaver-hat*, a beaver must have been *hunted* and killed, and, in the doing of which, many beavers may have been *wounded* and left to pine away the rest of their lives; and, perhaps many little orphan beavers, left to lament the murder of their parents. BEN LEY was the only real and sincere Pythagorean of modern times, that I ever heard of. He protested, not only against eating the flesh of animals, but also against robbing their backs; and, therefore, his dress consisted wholly of *flax*. But, even he, like Sir Richard Phillips, eat milk, butter, cheese, and eggs; though this was cruelly robbing the hens, cows, and calves; and, indeed causing the murder of the calves. In addition, poor little BEN forgot the materials of *book-binding*; and, it was well he did; for else, his Bible would have gone into the fire!

373. Taking it for granted, then, that sportsmen are as good as other folks on the score of *humanity*, the sports of the field, like every thing else done in the fields, tend to produce, or preserve *health*. I prefer them to all other pastime, because they produce *early rising*; because they have no tendency to lead young men into vicious habits. It is where men *con-*

gregate that the vices haunt. A hunter or a shooter may also be a gambler and a drinker; but, he is *less likely* to be fond of the two latter, if he be fond of the former. Boys will take to *something* in the way of pastime; and, it is better that they take to that which is innocent, healthy, and manly, than that which is vicious, unhealthy, and effeminate. Besides, the scenes of rural sport are necessarily at *a distance from cities and towns.* This is another great consideration; for though great talents are wanted to be *employed* in the *hives of men,* they are very rarely *acquired* in these hives: the surrounding objects are too numerous, too near the eye, too frequently under it, and too artificial.

374. For these reasons I have always encouraged my sons to pursue these sports. They have, until the age of 14 or 15, spent their time, by day, chiefly amongst horses and dogs, and in the fields and farm-yard; and their candle-light has been spent chiefly in reading books about hunting and shooting and about dogs and horses. I have supplied them plentifully with *books* and *prints* relating to these matters. They have *drawn* horses, dogs, and game themselves. These things, in which they took so deep an interest, not only engaged their attention and wholly kept them from all taste for,

and even all knowledge of, *cards* and other senseless amusements*;* but, they led them *to read and write of their own accord;* and, *never in my life have I set them a copy in writing nor attempted to teach them a word of reading.* They have learnt to read by looking into books about dogs and game; and they have learnt to write by imitating my writing, and by writing endless letters to me, when I have been from home, about their dogs and other rural concerns. While the Borough-tyrants had me in Newgate for two years, with a thousand pounds fine, for having expressed my indignation at their flogging of Englishmen, in the heart of England, under a guard of Hanoverian sabres, I received *volumes of letters* from my children; and, I have them now, from the *scrawl* of *three years*, to the neat and beautiful hand of thirteen. I never told them of any *errors* in their letters. All was well. The best evidence of the utility of their writing, and the strongest encouragement to write again, was *a very clear answer from me*, in a very precise hand, and upon very nice paper, which they never failed promptly to receive. They have all written to me *before they could form a single letter.* A little bit of paper, with some ink-marks on it, folded up by themselves, and a wafer stuck in it, used to be sent to me, and it was *sure* to bring the writer a very, very kind answer. Thus have they gone on. So far from being a *trouble* to me, they have been all *pleasure* and *advantage*. For many years they have been so many *secretaries*. I have *dictated* scores of registers to them, which have *gone to the press without my ever looking at them*. I dictated registers to them at the age of *thirteen*, and even of *twelve*. They have, as to *trust-worthiness*, been grown persons, at eleven or twelve. I could leave my house and affairs, the paying of men, or the going from home on business, to them at an age when boys in England, in general, want servants to watch them to see that they do not kill chickens, torment kittens, or set the buildings on fire.

375. Here is a good deal of *boasting*; but, it will not be denied, that I have *done a great deal* in a short public life, and I see no harm in telling my readers of any of the means, that I have employed; especially as I know of few greater misfortunes than that of breeding up things to be *school-boys all their lives*. It is not, that I have so many wonders of the world: it is that I have pursued a rational plan of education, and one that any man may pursue, if he will, with similar effects. I remembered, too, that I myself had had a sportsman-education. I ran after the hare-hounds at the age of *nine or*

ten. I have many and many a day left the rooks to dig up the wheat and peas, while I followed the hounds; and have returned home at dark-night, with my legs full of thorns and my belly empty to go supperless to bed, and to congratulate myself if I escaped a flogging. I was *sure* of these consequences; but that had not the smallest effect in restraining me. All the lectures, all the threats, vanished from my mind in a moment upon hearing the first cry of the hounds, at which my heart used to be ready to bound out of my body. I *remembered* all this. I traced to this taste my contempt for card-playing and for all childish and effeminate amusements. And, therefore, I resolved to leave the same course freely open to my sons. This is *my plan* of education: others may follow what plan they please.

376. This Chapter will be a head without a body; for, it will not require much time to give an account of the rural sports in America. The general taste of the country is to *kill* the things in order to have them to *eat*, which latter forms no part of the *sportsman*'s objects.

377. There cannot be said to be any thing here, which we, in England, call *hunting*. The deer are hunted by *dogs*, indeed, but the hunters do not *follow*. They are *posted* at their several stations to *shoot* the deer as he passes. This

is only one remove from the *Indian* hunting. I never saw, that I know of, any man that had *seen* a *pack of hounds* in America, except those kept by old JOHN BROWN, in Bucks County, Pennsylvania, who was the only *hunting Quaker* that I ever heard of, and who was grandfather of the famous General Brown. In short, there is none of what we call hunting; or, so little, that no man can expect to meet with it.

378. No *coursing*. I never saw a greyhound here. Indeed, there are no *hares* that have the same manners that ours have, or any thing like their fleetness. The woods, too, or some sort of cover, except in the singular instance of the *plains* in this Island, are too near at hand.

379. But, of *shooting* the variety is endless. Pheasants, partridges, wood-cocks, snipes, grouse, wild-ducks of many sorts, teal, plover, rabbits.

380. There is a disagreement between the North and the South as to the *naming* of the two former. North of New Jersey the pheasants are called partridges, and the partridges are called quails. To the South of New Jersey, they are called by what I think are their proper names, taking the English names of those birds to be proper. For, pheasants do not remain in *coveys;* but, mix, like common fowls. The intercourse between the males and females is

promiscuous, and not by *pairs*, as in the case of partridges. And these are the manners of the American pheasants, which are found by ones, twos, and so on, and never in *families*, except when *young*, when, like chickens, they keep with the old hen. The American *partridges* are not *quails;* because quails are *gregarious*. They keep in *flocks*, like *rooks* (called *crows* in America), or like *larks*, or *starlings;* of which the reader will remember a remarkable instance in the history of the migration of those grumbling vagabonds, the Jews, soon after their march from HOREB, when the quails came and settled upon each other's backs to a height of two cubits, and covered a superficial space of two days' journey in diameter. It is a well known fact, that quails *flock:* it is also well known, that partridges do not, but that they keep in *distinct families*, which we call *coveys* from the French *couvée*, which means the eggs or brood which a hen *covers* at one time. The American partridges live in coveys. The cock and her *pair* in the spring. They have their brood by *sitting alternately* on the eggs, just as the English partridges do; the young ones, if none are killed, or die, remain with the old ones till spring; the covey always live within a small distance of the same spot; if frightened into a state of separation, they *call* to each other and re-assemble; they roost all together in a round ring, as close as they can sit, the tails inward and the heads outward; and are, in short, in all their *manners*, precisely the same as the English partridge, with this exception, that they will sometimes alight on a rail or a bough, and that, when the hen sits, the cock, perched at a little distance, makes a sort of periodical whistle, in a monotonous, but very soft and sweet tone.

381. The size of the pheasant is about the *half* of that of the English. The plumage is by no means so beautiful; but, the flesh is far more delicate. The size of the partridge bears about the same proportion. But its plumage is more beautiful than that of the English, and its flesh is more delicate. Both are delightful, though rather difficult, shooting. The pheasant does not *tower*, but darts through the trees; and the partridge does not rise boldly, but darts away at no great height from the ground. Some years they are more abundant than other years. This is an abundant year. There are, perhaps, fifty coveys within half a mile of my house.

382. The *wood-cocks* are, in all respects, like those in England, except that they are only about three-fifths of the size. They *breed* here; and are in such numbers, that some men kill

twenty brace, or more in a day. Their haunts are in marshy places, or woods. The shooting of them lasts from the fourth of July till the *hardish frosts* come. The last we killed this year was killed on the 21*st of November*. So that here are *five months* of this sport; and pheasants and partridges are shot from September to April.

383. The *snipes* are called ***English snipes,*** which they resemble in all respects, and are found in great abundance in the usual haunts of snipes.

384. The *grouse* is precisely like the Scotch grouse. There is only here and there a place where they are found. But, they are, in those places, killed in great quantities in the fall of the year.

385. As to *wild ducks* and other water-fowl, which are come at by lying in wait, and killed most frequently swimming, or sitting, they are slaughtered in whole flocks. An American counts the cost of powder and shot. If he is *deliberate* in every thing else, this habit will hardly forsake him in the act of *shooting*. When the sentimental flesh-eaters hear the report of his gun, they may begin to pull out their white handkerchiefs; for death follows his pull of the trigger, with, perhaps, even more certainty than it used to follow the lancet of DOCTOR RUSH.

386. The PLOVER is a fine bird, and is found in great numbers upon the plains, and in the cultivated fields, of this Island, and at a mile from my house. Plovers are very *shy* and *wary;* but they have ingenious enemies to deal with. A waggon, or carriage of some sort, is made use of to approach them; and then they are easily killed.

387. ***Rabbits*** are very abundant in some places. They are killed by shooting; for all here is done with the *gun.* No reliance is placed upon a dog.

388. As to *game-laws* there are none, except those which appoint the *times* for killing. People go where they like, and, as to wild animals, shoot what they like. There is the Common Law, which forbids *trespass,* and the Statute Law, I believe, of "*malicious trespass,*" or trespass *after warning.* And these are more than enough; for nobody, that I ever hear of, *warns people off.* So that, as far as *shooting* goes, and that is the sport which is the most general favourite, there never was a more delightful country than this Island. The sky is so fair, the soil so dry, the cover so convenient, the game so abundant, and the people, go where you will, so civil, hospitable, and kind.

CHAP. XIII.

PAUPERS.

389. It is a subject of great exultation in the hireling newspapers of the Borough-villains, that "*poverty* and *poor-rates* have *found their* "*way* to America." As to the former it is literally true; for the *poverty* that is here has, almost the whole of it, *come from Europe;* but, the means of *keeping the poor* arise here upon the spot.

390. Great sums of money are raised in New York, Philadelphia, Boston, and other great sea-ports, for the maintenance of "*the poor;*" and, the Boroughmongers eagerly catch at the published *accounts* of this concern, and produce them as *proofs,* that misery is as great in America as it is under their iron rod. I will strip them of this pretext in a few minutes.

391. Let us take New York, for instance. It is notorious that, whatever may be the number of persons relieved by poor rates, the greater part of them are *Europeans,* who have come hither, at different periods and under circumstances of distress, different, of course, in de-

gree. There is, besides, a class of persons here of a description very peculiar; namely; the *free negroes*. Whatever may have been the motives, which led to their emancipation, it is very certain, that it has saddled the white people with a charge. These negroes are a disorderly, improvident set of beings; and, the paupers, *in the country*, consist *almost wholly* of them. Take out the *foreigners* and the *negroes*, and you will find, that the paupers of New York do not amount to a *hundredth part* of those of Liverpool, Bristol, Birmingham, or London, population for population. New York is a sea-port, and the only great sea port of a large district of country. All the disorderly crowd to it. It teems with emigrants; but, even there, a pauper, who is a white, *native American*, is a great rarity.

392. But, do the Borough-villains think, that the word *pauper* has the same meaning here that it has under their scorpion rod? A pauper under them means a man that is *able and willing to work*, and who *does work like a horse*; and who is *so taxed*, has so much of his earnings taken from him *by them* to pay the interest of *their* Debt and the pensions of themselves and their wives, children, and dependents, that he is actually starving and fainting *at his work*. This is what is meant by *a*

pauper in England. But, at New York, a pauper is, *generally*, a man who is unable, or, which is more frequently the case, unwilling to work; who is become debilitated from a vicious life; or, who, like boroughmongers and Priests, finds it more pleasant to live upon the labour of others than upon his own labour. A pauper in England is fed upon bones, garbage, refuse meat, and "*substitutes for bread.*" A pauper here expects, and has, as much flesh, fish and bread and cake as he can devour. How gladly would many a little tradesman, or even little farmer, in England, exchange his diet for that of a New York pauper!

393. Where there are *such paupers* as those in England, there are *beggars;* because, when they find, that they are *nearly starved* in the former character, they will try the latter in spite of all the *vagrant acts* that any hell-born Funding system can engender. And, who *ever* saw a beggar in America? "I have!" exclaims some spye of the Boroughmongers, who hopes to become a Boroughmonger himself. And so have I too. I have seen a *couple* since I have been on this Island; and of them I will speak presently. But there are *different sorts* of beggars too as well as of paupers. In England a beggar is a poor creature, with hardly rags (mere rags) sufficient to cover its nakedness, so

far even as common decency requires. A wretched mortal, the bare sight of whom would freeze the soul of an American within him. A dejected, broken down thing, that approaches you bare-headed, on one knee, with a trembling voice, with " pray bestow your charity, " for the Lord Jesus Christ's sake have compas- " sion on a poor soul;" and, if you toss a *half-penny* into his ragged hat, he exclaims in an extacy, " God Almighty bless your *honour!*" though you, perhaps, be but a shoe-black yourself. An American beggar, dressed very much like other people, walks up to you as boldly as if his pockets were crammed with money, and, with a half smile, that seems to say, he doubts of the propriety of his conduct, very civilly asks you, *if you can* HELP *him to a quarter of a dollar*. He mostly states the precise sum; and never sinks below *silver*. In short, there is *no begging*, properly so called. There is nothing that resembles English begging even in the most distant degree.

394. I have now been here *twenty months*, and I have been visited by only *two beggars*. The first was an *Englishman*, and what was more to me, a *Surrey* man too; a native of Croydon. He asked me if I could *help* him to a quarter of a dollar; for, it is surprising how apt scholars they are. " Yes," said I, " if you will

"*help* my men to do some work first." He said he could not do that, for he was *in a hurry*. I told him, that, if a man, with a dollar a day, and pork for the tenth part of a dollar a pound, could not earn his living, he ought to be hanged; "however," said I, "as you "are the first Surrey man I ever saw in Ame- "rica besides myself, if you be not hanged be- "fore this day week, and come here again, I "will *help* you to a quarter of a dollar." He came, and I kept my word. The second beggar was an *Italian*. This was a personage of "*high consideration*." He was introduced to the side of my writing table. He behaved with a sort of dignified politeness, mixed with somewhat of reserve, as if he thought the person to whom he was addressing himself a very good sort of man, but of rank inferior to himself. We could not understand each other at first; but, we got into *French*, and then we could talk. He having laid down his hat, and being seated, pulled out a large parcel of papers, amongst which was a certificate from the *Secretary of State of His Majesty the King cf Sardinia*, duly signed and countersigned, and sealed with a seal having the armorial bearings of that sovereign. Along with this respectable paper was an English translation of it, done at New York, and authenticated by the Mayor and a Notary Public, with all due formality. All the time these papers were opening, I was wondering what this gentleman could be. I read, and stared, and read again. I was struck not less by the novelty than the audacity of the thing. "So then," said I, breaking silence, "your sovereign, after taxing you to "your ruin, has been graciously pleased to "give you credentials to show, that he *autho-* "*rizes you to beg in America;* and, not only "for yourself but for *others;* so that you are "an accredited ambassador from the beggars "in Sardinia!" He found he was got into *wrong hands:* and endeavoured to put *an end* to the negociation at once, by observing, that I was not *forced* to give, and that my *simple negative* was *enough.* "I beg your pardon, "Sir," said I, "you have submitted your *case* "to me; you have made an *appeal* to me; your "statement contains reasons for my giving; and "that gives me a right to shew, if I can, why I "ought not to give." He then, in order to prevent all *reasoning*, opened his Subscription, or Begging-book, and said: "you see, Sir, others "give!" "Now," said I, "you reason, but "your reasoning is *defective;* for, if you were "to shew me, that you had robbed all my "neighbours without their resenting it, would "it follow that I must let you rob me too?' "*Ah! par bleu,*" said he, snatching up his "credentials, "*je vois que vous êtes un avare.*"

—*Ah! by Old Nick, I see you are a Miser.*— And off he went; not, however, before I had had time to tell him to be sure to give my best respects to the king of Sardinia, and to tell His Majesty to keep his beggars at home.

395. I afterwards found, that cases like this are by no means rare; and that, in Pennsylvania, in particular, they have accredited beggars from all parts of the continent of Europe. This may be no unuseful hint for the English Boroughmongers, who have an undoubted claim to precedence before the German and Italian beggars. The Boroughmongers may easily add a legation of mendicity to their Envoyships and Consulships, without any great disgrace to the latter; and, since they can get nothing out of America by bullying and attacking, try what can be gained by canting and begging. The chances are, however, that many of them will, before they die, be beggars in their own proper persons and for their own use and behoof; and thus give a complete rounding to their career; plunderers in prosperity, and beggars in adversity.

396. As to the *poor-rates*, the real poor-rates, you must look to *the country*. In England the poor-rates *equal in amount the rent of the land!* Here, I pay, in poor-rates, only *seven* dollars upon a rent of *six hundred!* And I pay my

full share. In short, how is it *possible*, that there should be paupers to any amount, where the common average wages of a labourer are six dollars a week; that is to say, *twenty-seven shillings sterling*, and where the necessaries of life are, upon an average, of *half* the price that they are in England? How can a man be a pauper, where he can earn ten pounds of prime hog-meat a day, six days in every week? I was at a *horse-race*, where I saw at least five thousand men, and not one man in shabby clothes.

397. But, some *go back* after they come from England; and the Consul at New York has thousands of applications from men who *want to go to Canada;* and little bands of them go off to that *fine country* very often. These are said to be *disappointed* people. Yes, they expected the people at New York to come out in boats, I suppose, carry them on shore, and give up their dinners and beds to them! If they will *work*, they will soon find beds and dinners: if they will not, they ought to have none. What, did they expect to find here the same faces and the same posts and trees that they left behind them? Such foolish people are not worth notice. The *lazy*, whether male or female, all hate a government, under which every one enjoys his earnings, *and no more*, Low,

poor and miserable as they may be, their *prin-ple* is precisely the same as that of Borough-mongers and Priests: namely, *to live without labour on the earnings of others.* The *desire* to live thus is almost universal; but with sluggards, thieves, Boroughmongers, and Priests, it is *a principle of action.* Ask a Priest *why* he is a Priest. He will say (for he has vowed it on the Altar!) that he believes himself called by the Holy Ghost to take on him the care of souls. But, put the thing close to him; push him hard; and you will find it was the *benefice,* the *money* and the *tithes,* that called him. Ask him what he wanted them for. That he might *live,* and live, too, *without work.* Oh! this work! It is an old saying, that, if the Devil find a fellow idle, he is sure to set him to work; a saying the truth of which the Priests seem to have done their utmost to establish.

398. Of the *goers back* was a Mr. ONSLOW WAKEFORD, who was a coach-maker, some years, in Philadelphia, and who, having, from nothing hardly to begin with, made a comfort-able fortune, *went back* about the time that I returned home. I met him, by accident, at Goodwood, in Sussex, in 1814. We talked about America. Said he, " I have often thought " *of the foolish way,* in which my good friend, " NORTH, and I used to talk about the happy

"state of England. *The money that I have* "*paid in taxes here, would have kept me like a* "*gentleman there.* Why," added he, "if a la-"bouring man here were seen *having in his* "*possession*, the fowls and other things that "labourers in Philadelphia carry home from "market, he would be stopped in the street, "and *taken up on suspicion of being a thief;* upon "the supposition of its being *impossible* that he "could have come *honestly* by them." I told this story after I got home; and we read in the news-papers, not long afterwards, that a *Scotch Porter*, in London, who had had a *little tub of butter* sent him up from his relations, and who was, in the evening, carrying it from the vessel to his home, had actually been seized by the Police, lodged in prison all night, brought before the magistrate the next day, and not released until he had produced witnesses *to prove* that he had *not stolen a thing, which was thought far too valuable for such a man to come at by honest means!* What a state of things must that be? What! A man in England taken up as a thief and crammed into prison, merely because he was in possession of 20 pounds of butter!

399. Mr. WAKEFORD is, I dare say, alive. He is a very worthy man. He lives at CHICHESTER. I appeal to him for the truth of the anecdote relating to him. As to the *butter*

story, I cannot name the *precise date;* but, I seriously declare the fact to have been as I have related it. I told Mr. WAKEFORD, who is a very *quiet* man, that, in order to make his lot in England as good as it was in America, he must help us to destroy the Boroughmongers. He left America, he told me, principally in consequence of the loss of his daughter (an only child) at Philadelphia, where she, amongst hundreds and hundreds of others, fell before the desolating *lancets* of 1797, 1798 and 1799.

CHAP. XIV.

GOVERNMENT, LAWS, AND RELIGION.

400. MR. PROFESSOR CHRISTIAN, who has written great piles of *Notes* on Blackstone's Commentaries, and whose Notes differ from those of the Note-writers on the Bible, in this, that the latter only tend to add darkness to that which was sufficiently dark before, while the Professor's Notes, in every instance, without a single exception, labour most arduously, and not always without success, to render that obscure, which was before clear as the sun now is in Long Island, on this most beautiful fifth of December, 1818: this Professor, who, I believe, is now a *Judge*, has, in his Note 126 on Book I, drawn what he calls "a *distinction*" between *Political* and *Civil* Liberty, which distinction contains as to ideas, manner, and expressions, a complete specimen of what, in such a case, a writer ought to avoid.

401. Leaving definitions of this sort to such conceited bunglers as the Professor, I will just give a *sketch* (for it can be nothing more) of the *Government* and *Laws* of this country.

402. The country is divided into *States.* Each of these States has its own separate government, consisting of a *Governor*, *Legislative Body*, and *Judiciary Department.* But, then there is a *General Government*, which is, in fact, the government of the whole *nation;* for, it alone can do any thing with regard to *other nations.* This General Government consists of a *President*, a *Senate*, a *House of Representatives*, all which together are called the *Congress.* The President is elected for *four years*, the Senate for *four years*, and the House of Representatives for *two years.*

403. In most of the State-Governments, the election is *annual* for the *House of Representatives.* In some the Governor and the Senate are elected for a longer period, not exceeding *four years* in any case. But, in some, the whole, Governor, Senate, and Representatives, are elected ANNUALLY; and this last appears now to be the *prevailing* taste.

404. The *suffrage*, or *qualification of electors*, is very various. In some States every free man; that is, every man who is not *bondman* or *slave*, has a vote. In others, the payment of *a tax* is required. In others, a man must be *worth a hundred pounds.* In Virginia a man must be a *freeholder.*

405. This may serve to show how little Mr.

JERRY BENTHAM, the new Mentor of the Westminster Telemachus, knows about the political part of the American governments. Jerry, whose great, and, indeed, *only* argument, in support of *annual parliaments* and *universal suffrage*, is, that *America* is so *happy* under *such a system*, has, if we were to *own him*, furnished our enemies with a complete answer; for, they have, in order to silence him, only to refer to the *facts* of his argument of happy experience. By *silencing* him, however, I do not mean, the stopping of his tongue, or pen; for nothing but mortality will ever do that. This everlasting babbler has aimed a sort of stiletto stroke at me; *for what* God knows, except it be to act a consistent part, by endeavouring to murder the man whom he has so frequently robbed, and whose facts and thoughts, though disguised and disgraced by the robber's quaint phraseology, constitute the better part of his book. Jerry, who was made a Reformer by PITT's *refusal to give him a contract to build a penitentiary, and to make him prime administrator of penance*, that is to say, Beggar-Whipper General, is a very proper person to be toasted by those, who have plotted and conspired against Major Cartwright. Mr. Brougham *praises* Jerry: that is enough!

406. In the *four New England States*, the

qualification was *a hundred pounds.* But, one of those States, CONNECTICUT, has, to her great honour, recently set an example worthy of the imitation of the other three. A new constitution has, during this year, been formed in that State, according to which all the elections are to be *annual;* and, as to the *suffrage,* I will give it in the words of the instrument itself: "Every male white citizen of the United States, "who shall have gained a settlement in this "state, attained the age of twenty-one years, "and resided in the town [that is *parish* in the "English meaning] in which he may offer him-"self to be admitted to the privilege of being "an elector, at least six months preceding, "*and have a freehold estate of the yearly value* "*of seven dollars in this State;*—OR, having "been *enrolled in the militia,* shall have per-"formed military duty therein for the term of "one year, next preceding the time he shall "offer himself for admission, or, being liable "thereto, shall have been, by authority of law, "altogether excused therefrom;—OR, shall "*have paid a State Tax* within the year next "preceding the time he shall present himself for "admission, and shall sustain a good moral "character, shall, on his taking the oath pre-"scribed, *be an elector.*"

407. And then, the proof of bad moral cha-

racter, is, "a conviction of *bribery*, *forgery*, "*perjury*, *duelling*, *fraudulent bankruptcy*, "*theft*, or other offences, for which an in-"famous punishment is inflicted." By *forgery* is not, of course, contemplated *puff-out* forgery; for that, as an act of *resistance of oppression*, is fully justifiable: it is not only not an immoral, but it is a *meritorious* act. The *forgery* here meant is forgery committed against honest men, who, when they "*promise to pay*," mean to pay, and do pay when called upon. "*Bribery*" is very properly set at the head of the disqualifications; but, what a nest of villains it would exclude in England! *White* men are mentioned, but, another clause, admits all the Blacks *now* free, though it shuts out future *comers* of that colour, or of the yellow hue; which is perfectly just; for, Connecticut is not to be the receptacle of those, whom other States may choose to release from slavery, seeing that she has now *no slaves of her own*.

408. Thus, then, this *new* Constitution; a constitution formed by the *steadiest* community in the whole world; a constitution dictated by the most *ample experience*, gives to the people, as to the *three branches* of the government (the *Governor*, *Senate*, and *Representatives)* precisely what we reformers in England ask as to only *one* branch out of the three. Whoever

has a freehold worth a guinea and a half a year, though he pay no tax, and though he be not enrolled in the militia, has a vote. Whoever *pays a tax*, though he be not enrolled in the militia, and have no freehold, has a vote. Whoever is enrolled in the militia, though he have no freehold and pay no tax, has a vote. So that nothing but beggars, paupers, and criminals, can easily be excluded; and, you will observe, if you please, Messieurs Boroughmongers, that the State taxes are *all direct*, and so contemptible in amount, as not to be, all taken together, enough to satisfy the maw of *a single sinecure place-man* in England; and that the Electors choose, and annually too, *King*, *Lords*, and *Commons*. Now, mind, this change has been deliberately made by the most deliberate people that ever lived on the earth. New England is called, and truly, "the Land of "*Steady Habits*;" but, a Connecticut man is said to be a "*full-blooded* Yankey," and Yankey means *New Englander*. So that, here are the *steadiest* of the *steady* adopting, after all their usual deliberation and precaution, in a time of profound tranquillity, and without any party spirit or delusion, the plan of us "*wild* and "*mad*" Reformers of Old England. Please God, I will, before I go home, perform a pilgrimage into this State!

409. In *Virginia*, and the States where negro slavery exists, the slaves are *reckoned amongst the population in apportioning the seats in the General Congress.* So that, the slaves do not vote; but, their *owners have votes for them.* This is what Davis Giddy, Wilberforce, and the Spawn of the Green Room, call *virtual representation.* And this, to be sure, is what Sir FRANCIS BURDETT, in his speech at the Reading Dinner, meant by *universal* INTERESTS! From *universal suffrage*, he came down to *general suffrage:* this was only *nonsense;* but, *universal* INTERESTS is downright borough-mongering. Well may he *despair* of doing any *good* in the House of Commons! "Universal *interests*" is the Virginian plan; and, *in that state of things*, by no means unwise or unjust; for, it is easier to *talk* about freeing black slaves, then it is to *do* it. The *planters* in the Southern States are not to blame for having slaves, until some man will show how they are to get rid of them. No one has yet discovered the means. *Virtual representation*, or, in other words, *Universal interests*, is as good a thing as any one can devise for those States; and, if SIR FRANCIS will but boldly declare, that the people of England *must necessarily remain slaves*, his joining of Davis Giddy and Canning, will be very consistent. Let him black the skins of the people of Eng-

land, and honestly call a part of them his property, and then he will not add the meanest to the most dastardly apostacy.

410. The right of suffrage in America is, however, upon the whole, sufficient to guard the people against any general and long-existing abuse of power; for, let it be borne in mind, that *here* the people elect *all* the persons, who are to exercise power; while, even if our Reform were obtained, there would still be *two branches out of the three*, over whom the people would have no direct controul. Besides, in England, Ireland, and Scotland, there is an *established Church;* a richly endowed and powerful hierarchy; and this, which is really a *fourth* branch of the government, has nothing to resemble it in America. So that, in this country, the whole of the Government may be truly said to be in the hands of the people. The people are, in reality as well as in name, *represented.*

411. The consequences of this are, 1st. that, if those who are chosen do not behave well, they are *not chosen a second time;* 2nd, that there are no *sinecure placemen* and *place women, grantees, pensioners without services,* and big placemen who swallow the earnings of two or three thousand men each; 3rd, that there is no military staff to devour more than the whole of a government ought to cost; 4th, that there are

no proud and insolent grasping Borough-mongers, who make the people toil and sweat to keep them and their families in luxury; 5th, that seats in the Congress are not like stalls in Smithfield, bought and sold, or hired out; 6th, that the Members of Congress do not sell their votes at so much a vote; 7th, that there is no waste of the public money, and no expenses occasioned by the bribing of electors, or by the hiring of Spies and informers; 8th, that there are no shootings of the people, and no legal murders committed, in order to defend the government against the just vengeance of an oppressed and insulted nation. But, all is harmony, peace and prosperity. Every man is zealous in defence of the laws, because every man knows that he is governed by laws, to which he has really and truly given his assent.

412. As to the nature of the Laws, the *Common Law* of England is the *Common Law* of America. These States were formerly *Colonies of England.* Our Boroughmongers wished to tax them *without their own consent.* But, the Colonies, standing upon the ancient Laws of England, which say that *no man shall be taxed without his own consent,* resisted the Boroughmongers of that day; overcame them in war; cast off all dependence, and became free and independent States. But, the great man, who con-

ducted that Revolution, as well as the people in general, were too wise to cast off the excellent laws of their forefathers. They, therefore, declared, that the *Common Law* of England should remain, being subject to such modifications as might be necessary in the new circumstances in which the people were placed. The *Common Law* means, the *ancient and ordinary usages and customs of the land* with regard to the means of *protecting property and persons* and of *punishing crimes.* This law is no *written* or *printed* thing. It is more ancient than books. It had its origin in the hearts of our forefathers, and it has lived in the hearts of their sons, from generation to generation. Hence it is emphatically called *the Law of the Land.* Juries, Judges, Courts of Justice, Sheriffs, Constables, Head-boroughs, Heywards, Justices of the Peace, and all their numerous and useful powers and authorities, make part of this *Law of the Land.* The Boroughmongers would fain persuade us, that it is *they* who have *given* us this Law, *out of pure generosity.* But, we should bear in mind, that this Law is more ancient, and far more ancient, than the titles of even the most ancient of their families. And, accordingly, when the present Royal Family were placed upon the throne, there was a solemn *declaration* by the Parliament in these words:

" The Laws of England are the *Birthright* of " the People of England." The Boroughmongers, by giving new powers to Justices of the Peace and Judges, setting aside the trial by Jury in many cases, both of property and person, even before the present horrible acts; and by a thousand other means, have, by *Acts of Parliament*, greatly despoiled us of the *Law of the Land;* but, never have they given us any one good in addition to it.

413. The Americans have taken special care to prevent the like encroachments on their rights: so that, while they have Courts of Justice, Juries, Judges, Sheriffs, and the rest, as we have; while they have all the *good* part of the Laws now in force in England, they have none of the *bad*. They have none of that *Statute Law* of England, or *Act of Parliament Law*, which has robbed us of a great part, and the *best* part of our " Birthright."

414. It is, as I said before, not my intention to go much into particulars here; but, I cannot refrain from noticing, that the People of America, when they come to settle their new governments, took special care to draw up specific *Constitutions*, in which they forbade any of their future law-makers to allow of any *Titles of Nobility*, any *Privileged Class*, any *Established Church*, or, to *pass any law* to give to

any body *the power of imprisoning men otherwise than in due course of Common Law*, except in cases of actual *invasion* or *open rebellion*. And, though actual invasion took place several times during the late war; though the Capital city was in possession of our troops, no such law was passed. Such is the effect of that confidence, which a good and just government has in the people whom it governs!

415. There is one more particular, as to the Laws of America, on which, as it is of very great importance, I think it right to remark. The uses, which have been made of the *Law of Libel* in England are well known. In the first place, the *Common Law* knows of no such offence as that of *criminal libel*, for which so many men have been so cruelly punished in England. The crime is an invention of late date. The Common Law punished men for *breaches of the peace*, but no *words*, whether written or spoken, can be a breach of the peace. But, then some Boroughmonger judges said, that words might *tend to produce* a breach of the peace; and that, therefore, it was *criminal* to use such words. This, though a palpable stretch of law, did, however, by usage, become law so far as to be acted upon in America as well as in England; and, when I lived in the

State of PENNSYLVANIA, eighteen years ago, the Chief Justice of that State, finding even this law not sufficiently large, gave it another stretch to make it fit me. Whether the Legislature of that State will repair this act of injustice and tyranny remains yet to be seen.

116. The State of NEW YORK, in which I now live, awakened, probably by the act of tyranny, to which I allude, has taken care, by an Act of the State, passed in 1805, to put an end to those attacks on the press by charges of *constructive libel*, or, at least, to make the law such, that no man shall suffer from the preferring of any such charges unjustly.

417. The principal effect of this twisting of the law was, that, whether the words published were *true* or *false* the *crime* of publishing was *the same;* because, whether true or false, they *tended to a breach of the peace!* Nay, there was a Boroughmonger Judge in England, who had laid it down as *law*, that the *truer* the words were, the *more criminal* was the libel; because, said he, a breach of the peace was more likely to be produced by telling *truth* of a *villain*, than by telling *falsehood* of a *virtuous man*. In point of fact, this was true enough, to be sure; but what an infamous doctrine! What a base, what an unjust mind must this man have had!

418. The State of New York, ashamed that there should any longer be room for such miserable quibbling; ashamed to leave the Liberty of the Press exposed to the changes and chances of a doctrine so hostile to common sense as well as to every principle of freedom, passed an *Act*, which makes the *truth* of any publication a *justification* of it, provided the publisher can shew, that the publication was made with *good motives* and *justifiable ends;* and who can *possibly* publish *truth* without being able to shew *good motives* and *justifiable ends?* To expose and censure tyranny, profligacy, fraud, hypocrisy, debauchery, drunkenness: indeed, all sorts of wickedness and folly; and to do this in the words of *truth*, must tend, *cannot fail to tend*, to check wickedness and folly, and to strengthen and promote virtue and wisdom; and these, and *these only*, are the uses of the press. I know it has been said, for I have heard it said, that this is going *too far;* that it would tend to lay open the *private affairs of families*. And what then? Wickedness and folly should meet their *due* measure of censure, or ridicule, be they found where they may. If the faults of private persons were too trifling to deserve public notice, the mention of them would give the parties no pain, and the publisher would be despised for his tittle-tattle;

that is all. And, if they were of a nature so grave as for the exposure of them to give the parties *pain*, the exposure would be *useful*, as a warning to others.

419. Amongst the persons whom I have heard express a wish, to see the press what they called *free*, and at the same time to extend the restraints on it, with regard to persons in their private life, *beyond the obligation of adherence to truth*, I have never, that I know of, met with one, who had not some *powerful motive of his own* for the wish, and who did not feel that he had some vulnerable part about himself. The common observation of these persons, is, that *public men are fair game*. Why *public* men only? Is it because *their* wickedness and folly affect the public? And, how long has it been, I should be glad to know, since bad example in private life has been thought of no consequence to the public? The press is called "the *guardian* of the *public* "*morals*;" but, if it is to meddle with none of the vices or follies of individuals in private life, how is it to act as the guardian of the morals of the whole community? A press perfectly free, reaches these vices, which the *law* cannot reach without putting too much power into the hands of the magistrate. Extinguish the press, and you must let the magistrate into every private

house. The experience of the world suggests this remark; for, look where you will, you will see virtue in all the walks of life hand in hand with freedom of discussion, and vice hand in hand with censorships and other laws to cramp the press. England, once so free, so virtuous and so happy, has seen misery and crimes increase and the criminal laws multiply in the exact proportion of the increase of the restraints on the press and of the increase of the severity in punishing what are called libels. And, if this had not taken place it would have been very wonderful. Men who have the handling of the public money, and who know that the parliament is such as to be *silenced*, will be very apt to squander that money; this squandering causes heavy taxes; these produce misery amongst the greater number of the people; this misery produces crimes; to check these new penal laws are passed. Thus it is in England, where new hanging places, new and enlarged jails, prisons on the water, new modes of transporting, a new species of peace officers, a new species of Justices of the Peace, troops employed regularly in aid of the magistrate, and at last, spies and blood-money bands, all proclaim a real revolution in the nature of the government. If the *press* had continued *free*, these sad effects of a waste of the public money never could have taken place; for, the wasters of that money would have been so exposed as to be unable to live under the odium which the exposure would have occasioned; and, if the parliament had not checked the waste and punished the wasters, the public indignation would have destroyed the parliament. But, with a muzzled press, the wasters proceeded with the consciousness of impunity. Say to any individual man when he is 20 years of age: "You shall "do just what you please with all the money of "other people that you can, by any means, all "your life long, get into your hands, and no one "shall ever be permitted to make you account- "able, or even to write or speak a word against "you for any act of fraud, oppression, or waste." Should you expect such an individual to act honestly and wisely? Yet, this, in fact, is what a Boroughmonger Parliament and the new Law of Libel say to every set of Ministers.

420. Before I quit this subject of *Libel*, let me observe, however, that *no juryman*, even as the law now stands *in England*, is *in conscience* bound to find any man guilty on a charge of *criminal* libel, unless the *evidence prove* that the pretended libeller has been actuated by an *evil motive*, and unless it be also *proved by evidence*, that his words, spoken or written, were *scandalous* and *malicious*. Unless these things be

clearly proved by evidence, the juryman, who finds a man *guilty*, is a base, *perjured villain;* and ought to be punished as such.

421. The State of Connecticut, in her new Constitution, before mentioned, has put this matter of libel on the true footing; namely; "In *all* prosecutions and indictments for *libel* "the TRUTH *may be given in evidence*, and the "*Jury* shall have the right to determine *the law* "*and the facts*." Thus, then, common sense has, at last, got the better; and TRUTH can, in this State, at least, in no case, be a *legal crime*. But, indeed, the press has NOW *no restraint* in America, other than that imposed by TRUTH. Men publish what they please, so long as they do not publish *falsehoods;* and, even in such cases, they are generally punished by the public contempt. The press is, therefore, *taken altogether*, what the magistrate always ought to be: "*a terror to evil doers*, and a *re*-"*ward to those who do well*." But, it is not *the name* of REPUBLIC that secures these, or any other of the blessings of freedom. As gross acts of tyranny may be committed, and as base corruption practised, under *that name* as under the name of *absolute monarchy*. And, it becomes the people of America to guard their minds against ever being, in any case, *amused with names*. It is *the fair representation*

of the people that is the cause of all the good; and, if this be obtained, I, for my part, will never quarrel with any body about *names*.

422. *Taxes* and *Priests;* for these always lay on heavily together. On the subject of taxes, I have, perhaps, spoken sufficiently clear before; but, it is a great subject. I will, on these subjects, address myself more immediately to *my old neighbours of Botley*, and endeavour to make them understand, what America is as to taxes and priests.

423. Worried, my old neighbours, as you are by tax-gatherers of all descriptions from the County-Collector, who rides in his coach and four down to the petty Window-Peeper, the little miserable spy, who is constantly on the look out for you, as if he were a thief-catcher and you were thieves; devoured as you are by these vermin, big and little, you will with difficulty form an idea of the state of America in this respect. It is a state of such blessedness, when compared with the state of things in England, that I despair of being able to make you fully comprehend what it is. Here a man may make new windows, or shut up old windows, as often as he pleases, without being compelled under a penalty to give notice to some insolent tax-gathering spy. Here he may keep as many horses as he likes, he may ride

them or drive them at his pleasure, he may sell them or keep them, he may lend them or breed from them; he may, as far as their nature allows, do the same with regard to his dogs; he may employ his servants in his house, in his stables, in his garden, or in his fields, just as he pleases; he may, if he be foolish enough, have armorial bearings on his carriage, his watch-seals, on his plate, and, if he likes, on his very buckets and porridge pots; he may write his receipts, his bills, his leases, his bonds, and deeds upon unstamped paper; his wife and daughters may wear French gloves and Lace and French and India silks; he may purchase or sell lands and may sue at law for his rights: and all these, and a hundred other things, without any dread of the interloping and insolent interference of a tax-gatherer or spy of any description. Lastly, when he dies, he can bequeath his money and goods and houses and lands to whomsoever he pleases; and he can close his eyes without curses in his heart against a rapacious band of placemen, pensioners, grantees, sinecure holders, staff-officers, borough-jobbers, and blood-money spies, who stand ready to take from his friends, his relations, his widow, and his children, a large part of what he leaves, under the name of a tax upon legacies.

424. But, you will ask, "are there *no taxes* "in America?" Yes; and taxes, or public contributions of some sort, there must be in every civilized state; otherwise *government* could not exist, and without government there could be no security for *property* or *persons*. The taxes in America consist principally of *custom duties imposed on goods imported into the country*. During the late war, there were taxes on several things in the country; but, they were taken off at the peace. In the cities and large towns, where *paving* and *lamps* and *drains* and *scavengers* are necessary, there are, of course, direct contributions to defray the expence of these. There are also, of course, *county rates* and *road rates*. But, as the money thus raised is employed for the immediate benefit of those who pay, and is expended amongst themselves and under their own immediate inspection, it does not partake of the nature of *a tax*. The taxes or duties, on goods imported, yield a great sum of money; and, owing to the persons employed in the collection being appointed for their integrity and ability, and not on account of their connection with any set of bribing and corrupt boroughmongers, the whole of the money thus collected is fairly applied to the public use, and is amply sufficient for all the purposes of government. The *army*, if it can be so called,

costs but a mere trifle. It consists of a few men, who are absolutely necessary to keep forts from crumbling down, and guns from rotting with rust. The navy is an object of care, and its support and increase a cause of considerable expence. But the government, relying on the good sense and valour of a people, who must hate or disregard themselves before they can hate or disregard that which so manifestly promotes their own happiness, has no need to expend much on any species of warlike preparations. The government could not stand a week, if it were hated by the people; nor, indeed, *ought* it to stand an hour. It has the hearts of the people with it, and, therefore, it need expend nothing in *blood-money*, or in *secret services* of any kind. Hence the *cheapness* of this government; hence the small amount of the *taxes*; hence the ease and happiness of the People.

425. Great as the distance between you and me is, my old neighbours, I very often think of you; and especially when I buy *salt*, which our neighbour Warner used to sell us for 19*s*. a bushel, and which I buy here for 2*s*. 6*d*. This salt is made, you know, down somewhere by Hambel. This very salt; when brought here from England, has all the charges of freight, insurance, wharfage, storage, to pay.

It pays besides, one third of its value in duty to the American Government before it be landed here. Then, you will observe, there is the *profit* of the American Salt Merchant, and then that of the shop-keeper who sells me the salt. And, after all this, I buy that very Hampshire salt for 2*s*. 6*d*. a bushel, English measure. What a government, then, must that of the Boroughmongers be! The salt is a gift of God. It is thrown on the shore. And yet, these tyrants will not suffer us to use it, until we have paid *them* 15*s*. a bushel for liberty to use it. They will not suffer us to use the salt, which God has sent us, until we have given them 15*s*. a bushel for them to bestow on themselves, on their families and dependants, in the payment of the interest of the Debt, which they have contracted, and in paying those, whom they hire to shoot at us. Yes; England is a fine country; it is a glorious country; it contains an ingenious, industrious, a brave and warm-hearted people; but, it is now disgraced and enslaved: it is trodden down by these tyrants; and we must free it. We cannot, and we will not die their slaves.

426. Salt is not the only one of the English articles that we buy cheaper here than in England. *Glass*, for instance, we buy for half the price that you buy it. The reason is, that you

are compelled to pay a *heavy tax*, which is not paid by us for that same glass. It is the same as to almost every thing that comes from England. You are compelled to pay the Boroughmongers a heavy tax on your *candles* and *soap.* You dare not *make* candles and soap, though you have the fat and the ashes in abundance. If you attempt to do this, you are taken up and imprisoned; and, if you resist, soldiers are brought to shoot you. This is *freedom*, is it? Now, we, *here*, make our own candles and soap. Farmers sometimes *sell* soap and candles; but they never *buy* any. A labouring man, or a mechanic, buys a sheep now and then. Three or four days' works will buy a labourer a sheep to weigh sixty pounds, with seven or eight pounds of loose fat. The meat keeps very well, in winter, for a long time. The wool makes stockings. And the loose fat is made into candles and soap. The year before I left Hampshire, a poor woman at Holly Hill had *dipped* some *rushes* in grease to use instead of candles. An Exciseman found it out; went and ransacked her house; and told her, that, if the rushes had had *another dip*, they would have been *candles*, and she must have gone to jail! Why, my friends, if such a thing were told here, nobody would believe it. The Americans could not bring their minds to believe, that Englishmen would submit to such atrocious, such degrading tyranny.

427. I have had living with me *an Englishman,* who smokes tobacco; and he tells me, that he can buy as much tobacco here for *three cents;* that is, about *three English half-pence,* as he could buy in England for *three shillings.* The *leather* has *no tax* on it here; so that, though the shoe-maker is paid a high price for his labour, the labouring man gets his shoes very cheap. In short, there is no *excise* here; no *property tax;* no *assessed taxes.* We have no such men here as Chiddel and Billy Tovery to come and take our money from us. No window peepers. No spies to keep a look-out as to our carriages and horses and dogs. Our dogs that came from Botley now run about free from the spying of tax-gatherers. We may wear hair-powder if we like without paying for it, and a boy in our houses may whet our knives without our paying two pounds a year for it.

428. But, then, we have not the honour of being covered over with the dust, kicked up by the horses and raised by the carriage-wheels of such men as Old GEORGE ROSE and OLD GARNIER, each of whom has pocketted more than *three hundred thousand pounds* of the public, that is to say, the people's, money. There are no such men here. Those who receive

public money here, do something for it. They *earn* it. They are no richer than other people. The *Judges* here are plain-dressed men. They go about with no sort of parade. They are dressed, on the Bench, like other men. The lawyers the same. Here are no black gowns and scarlet gowns and big foolish-looking wigs. Yet, in the whole world, there is not so well-behaved, so orderly, so steady a people; a people so *obedient to the law*. But, it is *the law only* that they will *bow* to. They will bow to nothing else. And, they bow with reverence to the law, because they know it to be just, and because it is made by men, whom they have all had a hand in choosing.

429. And, then, think of the *tithes!* I have talked to several farmers here about the tithes in England; and, they *laugh*. They sometimes almost make me angry; for they seem, at last, not to believe what I say, when I tell them, that the English farmer gives, and is compelled to give, the Parson a tenth part of his whole crop and of his fruit and milk and eggs and calves and lambs and pigs and wool and honey. They cannot believe this. They treat it as a sort of *romance*. I sometimes almost wish them to be farmers in England. I said to a neighbour the other day, in half anger: " I wish your farm were at Botley. There

" is a fellow there, who would soon let you " know, that your fine apple-trees do not belong " to you. He would have his nose in your " sheep-fold, your calf-pens, your milk-pail, " your sow's-bed, if not in the sow herself. " Your daughters would have no occasion to " hunt out the hen's nests: he would do that " for them." And then I gave him a proof of an English Parson's vigilance by telling him the story of Baker's peeping out *the name*, marked on the sack, which the old woman was wearing as a petticoat. To another of my neighbours, who is very proud of the circumstance of his grandfather being an *Englishman*, as, indeed, most of the Americans are, who are descended from Englishmen: to this neighbour I was telling the story about the poor woman at Holly Hill, who had nearly dipped her rushes once too often. He is a very grave and religious man. He looked very seriously at me, and said, that *falsehood* was *falsehood*, whether in jest or earnest. But, when I invited him to come to my house, and told him, that I would show him the acts which the Borough-men had made to put us in jail if we made our own soap and candles, he was quite astonished. " What! said he, and is *Old England really come to this!* Is the land of our " forefathers brought to this state of abject "slavery! Well, Mr. Cobbett, I confess, that "I was always for king George, during our "Revolutionary war; but, I believe, all was "for the best; for, if I had had my wishes, he "might have treated us as he now treats the "people of England." "*He!*" said I. "It is "not *he;* he, poor man, does nothing to the "people, and never has done any thing to the "people. *He* has no power more than you "have. None of his family have any. All "put together, they have not a thousandth part "so much as I have; for I am able, though "here, to annoy our tyrants, to make them "less easy than they would be; but, these "tyrants care no more for the Royal Family "than they do for so many posts or logs of "wood." And then I explained to him who and what the Boroughmongers were, and how they oppressed us and the king too. I told him how they disposed of the Church livings, and, in short, explained to him all their arts and all their cruelties. He was exceedingly shocked; but was glad, at any rate to know the *truth*.

430. When I was, last winter, in the neighbourhood of Harrisburgh in Pennsylvania, I saw some *hop-planters*. They grow prodigious quantities of hops. They are obliged to put their hills so wide a part, that they can have

only four hundred hills upon an acre; and yet they grow three thousand pounds of hops upon an acre, with no *manure* and with once ploughing in the year. When I told them about the price of hops in England and about the difficulty of raising them, they were greatly surprised; but, what was their astonishment, when I told them about the hop-poles of CHALCRAFT at Curbridge! The hop is naturally a *weed* in England as well as in America. Two or three vines had come up out of Chalcraft's garden hedge, a few years ago. Chalcraft put *poles* to them; and, there might be a pound or two of hops on these poles. Just before the time of gathering, one of the spies called *Excisemen* called on Chalcraft and asked him why he did not *enter* his hops. Chalcraft did not understand; but, answered, he meant to *take them in* shortly, though he did not think they were yet quite ripe. "Aye," said the Exciseman, "but I "mean, when do you mean to enter them at "the *excise office?*" Chalcraft did not know (not living in a hop-country,) that he had already incurred a *penalty* for not reporting to the tyrants that he had hops growing in his garden hedge! He did not know, that he could not gather them and put them by without giving notice, under a *penalty of fifty pounds*. He did not know, that he could not receive this little gift of God without paying money to the Borough-mongers in the shape of tax; and, to the Parson in the shape of tithe, or, to give a tenth of the hops to the Parson, and not dare pick a single hop till he had sent *notice to the Parson!* What he did, upon this occasion, I have forgotten; but, it is likely that he let the hops stand and rot, or cut them down and flung them away as weeds. Now, poor men in England are told to be *content* with rags and hungry bellies, for that is *their lot;* that " it has pleased Di-" vine *Providence* to place them in that state." But, here is a striking instance of the falsehood and blasphemy of this Doctrine; for, providence had sent Chalcraft the hops, and he had put poles to them. Providence had brought the hops to perfection; but then came the Boroughmongers and the Parson to take from this poor man this boon of a benevolent Maker. What, did God order a tax with all its vexatious regulations, to be imposed upon what he had freely given to this poor man? Did God ordain that, in addition to this tax, a *tenth* should be yielded to a Parson, who had solemnly vowed at his ordination, that he believed himself called, not by the love of tithes, but by " *the Holy Ghost,* to take on him the *cure of* " *souls,*" and to " *bring stray sheep into the* " *fold of the Lord?*" Did God ordain these things? Had it *pleased God* to do this? What impunity, what blasphemy, then, to ascribe to Providence the manifold sufferings occasioned by the Boroughmongers' taxes and Parson's tithes!

431. But, my Botley neighbours, you will exclaim, "No *tithes!* Why, then, there can "be no *Churches* and no *Parsons!* The people "must know nothing of God or Devil; and "must all go to hell!" By no means, my friends. Here are plenty of Churches. No less than three Episcopal (or English) Churches; three Presbyterian Churches; three Lutheran Churches; one or two Quaker Meeting-houses; and two Methodist Places; all within *six miles* of the spot where I am sitting. And, these, mind, not poor shabby Churches; but each of them larger and better built and far handsomer than Botley Church, with the Church-yards all kept in the neatest order, with a head-stone to almost every grave. As to the Quaker Meeting-house, it would take Botley Church into its belly, if you were first to knock off the steeple.

432. Oh, no! Tithes are not necessary to promote *religion.* When our Parsons, such as Baker, talk about *religion,* or *the church,* being in danger; they mean, that the *tithes* are in danger. They mean, that they are in dan-

ger of being compelled to work for their bread. This is what they mean. You remember, that, at our last meeting at Winchester, they proposed for us to tell the Prince Regent, that we would *support the Church.* I moved, to leave out the word *church,* and insert the word *tithes;* for, as there were many presbyterians and other dissenters present, they could not, with clear consciences, pledge themselves to support *the church.* This made them *furious.* It was lifting up the *mask;* and the parsons were enraged beyond measure.

433. Oh, no! *Tithes* do not mean *religion.* Religion means *a reverence for God.* And, what has this to do with tithes? Why cannot you reverence God, without Baker and his wife and children eating up a tenth part of the corn and milk and eggs and lambs and pigs and calves that are produced in Botley parish? The Parsons, in this country, are supported by those who choose to employ them. A man belongs to what congregation he pleases. He pays what is required by the rules of the congregation. And, if he think that it is not necessary for him to belong to any congregation, he pays nothing at all. And, the consequence is, that all is harmony and good neighbourhood. Here are not disputes about religion; or, if there be, they make no noise. Here is

no ill-will on this account. A man is never asked what religion he is of, or whether he be of any religion at all. It is a matter that nobody interferes in. What need, therefore, is there of an *established* Church. What need is there of tithes? And, why should not that species of property be taken for *public use?* That is to say, as far as it has any thing to do with religion? I know very well, that tithes do not operate as many people pretend; I know that those who complain most about them have the least right to complain; but, for my present purpose, it is sufficient to shew, that they have nothing to do with *religion.*

434. If, indeed, the Americans were wicked, disorderly, criminal people, and, of course, a miserable and foolish people: then we might doubt upon the subject: then we might possibly suppose, that their wickedness and misery arose, in some degree, at least, from the *want of tithes.* But, the contrary is the fact. They are the most orderly, sensible, and least criminal people in the whole world. A common labouring man has the feelings of a man of honour; he never thinks of violating the laws; he crawls to nobody; he will call every man *Sir*, but he will call no man *master*. When he utters words of respect towards any one, they do not proceed from fear or hope, but

from civility and sincerity. A native American labourer is never *rude* towards his employer, but he is never *cringing*.

435. However, the best proof of the inutility of an established Church is the absence of *crimes* in this country, compared to the state of England in that respect. There have not been three *felonies* tried in this country since I arrived in it. The Court-house is at two miles from me. An Irishman was tried for forgery in the summer of 1817, and the whole country was alive to go and witness the novelty. I have not heard of a man being hanged in the whole of the United States since my arrival. The Boroughmongers, in answer to statements like these, say that this is a *thinly inhabited* country. This very country is *more thickly settled than Hampshire.* The adjoining country, towards the city of New York is much more thickly settled than Hampshire. New York itself and its immediate environs contain nearly two hundred thousand inhabitants, and after London, is, perhaps, the first commercial and maritime city in the world. Thousands of sailors, ship-carpenters, dock-yard people, dray-men, boat-men, crowd its wharfs and quays. Yet, never do we hear of hanging; scarcely ever of a robbery; men go to bed with scarcely locking their doors; and

never is there seen in the streets what is called in England, *a girl of the town;* and, what is still more, never is there seen in those streets a *beggar*. I wish you, my old neighbours, could see this city of New York. Portsmouth and Gosport, taken together, are miserable holes compared to it. Man's imagination can fancy nothing so beautiful as its bay and port, from which two immense rivers sweep up on the sides of the point of land, on which the city is. These rivers are continually covered with vessels of various sizes bringing the produce of the land, while the bay is scarcely less covered with ships going in and out from all parts of the world. The city itself is a scene of opulence and industry: riches without insolence, and labour without grudging.

436. What Englishman can contemplate this brilliant sight without feeling some little pride that this city bears an English name? But, thoughts of more importance ought to fill his mind. He ought to contrast the ease, the happiness, the absence of crime which prevail here with the incessant anxieties, the miseries and murderous works in England. In his search after causes he will find them no where but in the *government:* and, as to an established church, if he find no sound argument to prove it to be an evil; at the very least he must conclude, that

it is *not a good;* and, of course that property to the amount of five millions a year is very unjustly as well as unwisely bestowed on its clergy.

437. Nor, let it be said, that the people here are of a better natural disposition than the people of England are. How can it be? They are, the far greater part of them, the immediate descendants of Englishmen, Irishmen, and Scotsmen. Nay, in the *city* of New York it is supposed, that a full half of the labour is performed by natives of Ireland, while men of that Island make a great figure in trade, at the bar, and in all the various pursuits of life. They have their Romish Chapels there in great brilliancy; and they enjoy "Catholic Emancipa-"tion" without any petitioning or any wrangling. In short, blindfold an Englishman and convey him to New York, unbind his eyes, and he will think himself in an English city. The same sort of streets; shops precisely the same; the same beautiful and modest women crowding in and out of them; the same play-houses; the same men, same dress, same language: he will miss by day only the nobility and the beggars, and by night only the street-walkers and pickpockets. These are to be found only where there is an *established* clergy, upheld by

what is called the *state*, and which word means, in England, *the Boroughmongers.*

438. Away, then, my friends, with all cant about *the church*, and the church being *in danger.* If the church, that is to say, the *tithes*, were completely *abolished;* if they, and all the immense property of the church, were taken and applied to public use, there would not be a sermon or a prayer the less. Not only the Bible but the very Prayer-book is in use here as much as in England, and, I believe, a great deal *more.* Why give the five millions a year then, to Parsons and their wives and children? Since the English, Irish, and Scotch, are so good, so religious, and so moral *here* without glebes and tithes; why not use these glebes and tithes for other purposes seeing they are possessions which can legally be disposed of in another manner?

439. But, the fact is, that it is the circumstance of the church being *established by law* that makes it *of little use* as to real religion, and as to morals, as far as they be connected with religion. Because, as we shall presently see, this establishment forces upon the people, parsons whom they cannot respect, and whom indeed, they must *despise;* and, it is easy to conceive, that the moral precepts of those, whom

we despise on account of their immorality, we shall never much attend to, even supposing the precepts themselves to be good. If a precept be self-evidently good; if it be an obvious duty which the parson inculcates, the inculcation is useless to us, because, whenever it is wanted to guide us, it will occur without the suggestion of any one; and, if the precept be not self-evidently good, we shall never receive it as such from the lips of a man, whose character and life tell us we ought to suspect the truth of every thing he utters. When the matters as to which we are receiving instructions are, in their nature, wholly dissimilar to those as to which we have witnessed the conduct of the teacher, we may reasonably, in listening to the precept, disregard that conduct. Because, for instance, a man, though a very indifferent Christian, may be a most able soldier, seaman, physician, lawyer, or almost any thing else; and what is more, may be honest and zealous in the discharge of his duty in any of these several capacities. But, when the conduct, which we have observed in the teacher belongs to the same department of life as the precept which he is delivering, if the one differ from the other we cannot believe the teacher to be sincere, unless he, while he enforces his precept upon us, acknow ledge his own misconduct. Suppose me, for instance, to be a great liar, as great a liar, if possible, as STEWART of the COURIER, who has said that I have been "fined 700 dollars "for writing against the American government," though I never was prosecuted in America in all my life. Suppose me to be as great a liar as STEWART, and I were to be told by a parson, whom I knew to be as great a liar as myself, that I should certainly go to hell if I did not leave off lying. Would his words have any effect upon me? No: because I should conclude, that if he thought what he said, he would not be such a liar himself. I should rely upon the parson generally, or I should not. If I did, I should think myself safe until I out-lied him; and, if I did not rely on him generally, of what use would he be to me?

440. Thus, then, if men be *sincere* about religion; if it be not all a mere matter of form, it must always be of the greatest consequence that the example of the teacher correspond with his teaching. And the most likely way to insure this, is to manage things so that he may in the first place, be selected by the people, and, in the second place, have no rewards in view other than those which are to be given in consequence of his perseverance in a line of good conduct.

441. And thus it is with the clergy in Ame-

rica, who are duly and amply rewarded for their diligence, and very justly respected for the piety, talent, and zeal which they discover; but, who have no tenure of their places other than that of the will of the congregation. Hence it rarely indeed happens, that there is seen amongst them an impious, an immoral, or a despicable man. Whether the teaching of even these Reverend persons have any very great *effect* in producing virtue and happiness amongst men is a question upon which men may, without deserving to be burnt alive, take the liberty to differ; especially since the world has constantly before its eyes a society, who excel in all the Christian virtues, who practise that simplicity which others teach, who, in the great work of charity, really and truly hide from the left hand that which the right hand doeth; and who know nothing of Bishop, Priest, Deacon, or Teacher of any description. Yes, since we have the Quakers constantly before our eyes, we may, without deserving to be burnt alive, question the utility of paying any parsons or religious teachers at all. But, the worst of it is, we are apt to confound things; as we have, by a figure of speech, got to call a *building a church*, when a church really means a body of people; so we are apt to look upon the *priest as being religious*, and especially when we call him *the reverend;* and, it often sadly occurs that no two things can be wider from each other in this quality. Some writer has said, that he would willingly leave to the clergy every thing above the tops of the chimneys; which, perhaps, was making their possessions rather too ethereal; but, since our law calls them "*spiritual* "*persons;*" since they profess, that "their king-"dom is not of this world," and, since those of our church have solemnly declared, that they believed themselves to be called to the ministry "by the Holy Ghost:" it is, I think, a little out of character for them to come poking and grunting and grumbling about after our eggs, potatoes, and sucking pigs.

442. However, upon the general question of the utility or non-utility of paid religious teachers, let men decide for themselves; but if teachers be to be paid, it seems a clear point, in my mind, that they should be paid upon the American plan: and this, I think, must be obvious to every one, who is able to take a view of the English Clergy. They are appointed by the absolute will of the Boroughmongers. They care nothing for the good will of their congregation or parish. It is as good to them to be hated by their parishioners as to be loved by them. They very frequently never even *see* their parish more than once in four or five years.

They solemnly declare at the altar, that they believe themselves called by the Holy Ghost to take on them the *cure of souls;* they get possession of a living; and leave the cure of souls to some *curate*, to whom they give a tenth part, perhaps, of the income. Many of them have *two livings*, at thirty miles distance from each other. They live at neither very frequently; and, when they do they only add to the annoyance which their curate gives.

443. As to their general character and conduct; in what public transaction of pre-eminent scandal have they not taken a part? Who were found most intimate with Mrs. CLARKE, and most busy in her commission dealing affairs? Clergymen of the Church of England. This is notorious. Miss TOCKER tells of the *two livings* given to PARSON GURNEY for his electioneering works in Cornwall. And, indeed all over the country, they have been and are the prime agents of the Boroughmongers. Recently they have been the tools of Sidmouth for gagging the press in the country parts of the kingdom. *Powis* and *Guillim* were the prosecutors of Messrs. Pilling and Melor; and for which if they be not made to answer, the kingdom ought to be destroyed. They are the leading men at Pitt Clubs all over the country; they were the foremost to defend the peculation of Melville. In short, there

has been no public man guilty of an infamous act, of whom they have not taken the part; and no act of tyranny of which they have not been the eulogists and the principal instrument.

444. But, why do I attempt to describe Parsons to *Hampshire men?* You saw them all assembled in grand cohort the last time that I saw any of you. You saw them at *Winchester*, when they brought forward their lying address to the Regent. You saw them on that day, and so did I; and in them I saw a band of more complete blackguards than I ever before saw in all my life. I then saw Parson Baines of Exton, standing up in a chair and actually spitting in Lord Cochrane's poll, while the latter was bending his neck out to speak. Lord Cochrane looked round and said, "B. "G— Sir, if you do that again I'll knock you "down." "You be d—d," said Baines, "I'll "spit where I like." Lord Cochrane struck at him; Baines jumped down, put his two hands to his mouth in a huntsman-like way, and cried "whoop! whoop!" till he was actually black in the face. One of them trampled upon my heel as I was speaking. I looked round, and begged him to leave off. "You be d—d," said he, "you be d—d, Jacobin." He then tried to press on me, to stifle my voice, till I clapped my elbow into his ribs and made "the spiritual

"person" hiccup. There were about twenty of them mounted upon a large table in the room; and there they jumped, stamped, hallooed, roared, thumped with canes and umbrellas, squalled, whistled, and made all sorts of noises. As Lord Cochrane and I were going back to London, he said that, so many years as he had been in the navy, he never had seen a band of such complete blackguards. And I said the same for the army. And, I declare, that, in the whole course of my life, I have never seen any men, drunk or sober, behave in so infamous a manner. Mr. PHILLIPS, of Eling, (now Doctor Phillips) whom I saw standing in the room, I tapped on the shoulder, and asked, whether he was not ashamed. Mr. LEE, of the College; Mr. OGLE, of Bishop's Waltham; and DOCTOR HILL, of Southampton: these were exceptions. Perhaps there might be some others; but the *mass* was the most audacious, foul, and atrocious body of men I ever saw. We had done nothing to offend them. We had proposed nothing to offend them in the smallest degree. But, they were afraid of our *speeches:* they knew they could not answer us; and they were resolved, that, if possible, we should not be heard. There was one parson, who had his mouth within a foot of Lord Cochrane's ear, all the time his Lordship was speaking, and who kept on saying: "*You lie! you* "*lie! you lie! you lie!*" as loud as he could utter the words.

445. BAKER, the Botley Parson, was extremely busy. He acted the part of buffoon to LOCKHART. He kept capering about behind him, and really seemed like a merry andrew rather than a "*spiritual person.*"

446. Such is the character of the great body of Hampshire Parsons. I know of no body of men so despicable, and yet, what sums of public money do they swallow! It now remains for me to speak more particularly of BAKER, he who, for your sins I suppose, is fastened upon you as your Parson. But what I have to say of this man must be the subject of another Letter. That it should be the subject of any letter at all may well surprize all who know the man; for not one creature knows him without despising him. But, it is not BAKER, it is the scandalous priest, that I strike at. It is the impudent, profligate, hardened priest that I will hold up to public scorn.

447. When I see the good and kind people here going to church to listen to some decent man of good moral character and of sober quiet life, I always think of you. You are just the same sort of people as they are here; but, what a difference in the Clergyman! What a differ-

ence between the sober, sedate, friendly man who preaches to one of these congregations, and the greedy, chattering, lying, backbiting, mischief-making, everlasting plague, that you go to hear, and are *compelled* to hear, or stay away from the church. Baker always puts me in mind of the Magpie.

> The Magpie, bird of chatt'ring fame,
> Whose tongue and hue bespeak his name;
> The first a *squalling clam'rous clack,*
> The last made up of white and black;
> Feeder alike on *flesh* and *corn,*
> Greedy alike at eve and morn;
> Of all the birds the *prying pest,*
> Must needs be *Parson* o'er the rest.

448. Thus I began a fable, when I lived at Botley. I have forgotten the rest of it. It will please you to hear that there are *no Magpies* in America; but, it will please you still more to hear, that no men that resemble them are parsons here. I have sometimes been half tempted to believe, that the Magpie first suggested to tyrants the idea of having a tithe-eating Clergy. The Magpie devours the corn and grain; so does the Parson. The Magpie takes the wool from the sheep's backs; so does the Parson. The Magpie devours alike the young animals and the eggs; so does the Parson. The Magpie's clack is everlastingly going; so is the

Parson's. The Magpie repeats by rote words that are taught it; so does the Parson. The Magpie is always skipping and hopping and peeping into other's nests: so is the Parson. The Magpie's colour is partly black and partly white; so is the Parson's. The Magpie's greediness, impudence, and cruelty are proverbial; so are those of the Parson. I was saying to a farmer the other day, that if the Boroughmongers had a mind to ruin America, they would another time, send over five or six good large flocks of Magpies, instead of five or six of their armies; but, upon second thought, they would do the thing far more effectually by sending over five or six flocks of their Parsons, and getting the people to receive them and cherish them as the *Bulwark* of *religion.*

END OF PART II.

A
YEAR'S RESIDENCE,
IN THE
UNITED STATES OF AMERICA.

Treating of the Face of the Country, the Climate, the Soil, the Products, the Mode of Cultivating the Land, the Prices of Land, of Labour, of Food, of Raiment; of the Expenses of Housekeeping, and of the usual manner of Living; of the Manners, Customs, and Character of the People; and of the Government, Laws, and Religion.

IN THREE PARTS.

By WILLIAM COBBETT.

PART III.

Containing,—Mr. Hulme's Introduction to his Journal—Mr. Hulme's Journal, made during a Tour in the Western Countries of America, in which Tour he visited Mr. Birkbeck's Settlement—Mr. Cobbett's Letters to Mr. Birkbeck, remonstrating with that Gentleman on the numerous delusions, contained in his two publications, entitled "Notes on a Journey in America" and "Letters from Illinois"—Postscript, being the detail of an experiment made in the cultivation of the Ruta Baga—Second Postscript, a Refutation of Fearon's Falsehoods.

LONDON:

PRINTED FOR SHERWOOD, NEELY, AND JONES,

PATERNOSTER-ROW.

1819.

CONTENTS OF PART III.

DEDICATION

To TIMOTHY BROWN, Esq.

OF PECKHAM LODGE, SURREY.

North Hempstead, Long Island,
10 *Dec.* 1818.

My dear Sir,

The little volume here presented to the public, consists, as you will perceive, for the greater and most valuable part, of travelling notes, made by our friend Hulme, whom I had the honour to introduce to you in 1816, and with whom you were so much pleased.

His activity, which nothing can benumb, his zeal against the twin monster, tyranny and priestcraft, which nothing can cool, and his desire to assist in providing a place of retreat for the oppressed, which nothing but success in the accomplishment can satisfy; these have induced him to employ almost the whole of his time here in various ways all tending to the same point.

The Boroughmongers have agents and spies all over the inhabited globe. Here they cannot *sell blood:* they can only collect information and calumniate the people of both countries. These vermin our friend *firks out* (as the Hampshire people call it); and they hate him as rats hate a terrier.

Amongst his other labours, he has performed a very laborious journey to the *Western Countries*, and has been as far as the Colony of our friend BIRKBECK. This journey has produced a JOURNAL; and this Journal, along with the rest of the volume, I dedicate to you in testimony of my constant remembrance of the many, many happy hours I have spent with you, and of the numerous acts of kindness, which I have received at your hands. You were one of those, who *sought acquaintance with me*, when I was shut up in a felon's jail *for two years* for having expressed my indignation at seeing Englishmen flogged, in the heart of England, under a guard of German bayonets and sabres, and when I had on my head *a thousand pounds fine* and *seven years' recognizances*. You, at the end of the two years, took me from the prison, in your carriage, home to your house. You and our kind friend, WALKER, are, *even yet*, held in bonds for my *good behaviour*, the seven years not being expired. All these things are written in the very core of my heart; and when I act as if I had forgotten any one of them, may no name on earth be so much detested and despised as that of

Your faithful friend,

And most obedient servant,

WM. COBBETT.

PREFACE TO PART III.

849. In giving an account of the United States of America, it would not have been proper to omit saying something of the *Western Countries*, that Newest of the New Worlds, to which so many thousands and hundreds of thousands are flocking, and towards which the writings of Mr. Birkbeck have, of late, drawn the pointed attention of all those Englishmen, who, having something left to be robbed of, and wishing to preserve it, are looking towards America as a place of refuge from the Boroughmongers and the Holy Alliance, which latter, to make the compact complete, seems to want nothing but the accession of His Satanic Majesty.

850. I *could not go* to the Western Countries; and, the accounts of others were seldom to be relied on; because, scarcely any man goes thither without some degree of partiality, or comes back without being tainted with some little matter, at least, of self-interest. Yet, it was desirable to make an attempt, at least, towards settling the question: "Whether the "Atlantic, or the Western, Countries were the "best for *English Farmers* to settle in." Therefore, when Mr. Hulme proposed to make a Western Tour, I was very much pleased, seeing that, of all the men I knew, he was the

most likely to bring us back an *impartial* account of what he should see. His great knowledge of farming as well as of manufacturing affairs; his capacity of estimating local advantages and disadvantages; the natural turn of his mind for discovering the means of applying to the use of man all that is furnished by the earth, the air, and water; the patience and perseverance with which he pursues all his inquiries; the urbanity of his manners, which opens to him all the sources of information: his inflexible adherence to *truth:* all these marked him out as the man, on whom the public might safely rely.

851. I, therefore, give his Journal, made during his tour. He offers no *opinion* as to the *question* above stated. That *I shall* do; and, when the reader has gone through the Journal he will find my opinions as to that question, which opinions I have stated in a Letter, addressed to Mr. BIRKBECK.

852. The American reader will perceive, that this Letter is intended principally for the perusal of *Englishmen;* and, therefore, he must not be surprised if he find a little bickering in a group so much of a *family* cast.

WM. COBBETT.

North Hempstead,
10*th December,* 1818.

A

YEAR'S RESIDENCE,

&c.

INTRODUCTION TO THE JOURNAL.

Philadelphia, 30th Sept. 1818.

853. It seems necessary, by way of Introduction to the following *Journal*, to say some little matter respecting the author of it, and also respecting his motives for wishing it to be published.

854. As to the first, I am an Englishman by birth and parentage; and am of the county of Lancaster. I was bred and brought up at farming work, and became an apprentice to the business of *Bleacher*, at the age of 14 years. My own industry made me a master-bleacher, in which state I lived many years at Great Lever, near Bolton, where I employed about 140 men, women, and children, and had generally about 40 apprentices. By this business, pursued with incessant application, I had ac-

quired, several years ago, property to an amount sufficient to satisfy any man of moderate desires.

855. But, along with my money my children had come and had gone on increasing to the number of *nine*. New *duties* now arose, and demanded my best attention. It was not sufficient that I was likely to have a decent fortune for each child. I was bound to provide, if possible, against my children being stripped of what I had earned for them. I, therefore, looked seriously at the situation of England; and, I saw, that the incomes of my children were all *pawned* (as my friend Cobbett truly calls it) to pay the Debts of the Borough, or seat, owners. I saw, that, of whatever I might be able to give to my children, as well as of what they might be able to earn, *more than one half* would be taken away to feed pensioned Lords and Ladies, Soldiers to shoot at us, Parsons to persecute us, and Fundholders, who had lent their money to be applied to purposes of enslaving us. This view of the matter was sufficient to induce the father of nine children to think of the means of rescuing them from the consequences, which common sense taught him to apprehend. But, there were other considerations, which operated with me in producing my emigration to America.

856. In the year 1811 and 1812 the part of the country, in which I lived, was placed under a *new sort of law;* or, in other words, it was placed out of the protection of the old law of the land. Men were seized, dragged to prison, treated like convicts, many transported and put to death, without having committed any thing, which the law of the land deems a *crime.* It was then that the infamous *Spy-System* was again set to work in Lancashire, in which horrid system FLETCHER of Bolton was one of the principal actors, or, rather, organizers and promoters. At this time I endeavoured to detect the machinations of these dealers in human blood; and, I narrowly escaped being sacrificed myself on the testimony of two men, who had their pardon offered them on condition of thei *swearing against me.* The men refused, and were transported, leaving wives and children to starve.

857. Upon this occasion, my friend DOCTOR TAYLOR, most humanely, and with his usual zeal and talent, laboured to counteract the works of FLETCHER and his associates. The DOCTOR published a pamphlet on the subject, in 1812, which every Englishman should read. I, as far as I was able, co-operated with him. We went to London, laid the real facts before several members of the two houses of Parlia-

ment; and, in some degree, checked the progress of the dealers in blood. I had an interview with Lord Holland, and told him, that, if he would pledge himself to cause the *secret-service money* to be kept in London, I would pledge myself for the keeping of the peace in Lancashire. In short, it was necessary, in order to support the tyranny of the seat-sellers, that *terror* should prevail in the populous districts. *Blood* was wanted to flow; and *money* was given to spies to tempt men into what the new law had made crimes.

858. From this time I resolved *not to leave my children in such a state of things*, unless I should be taken off very suddenly. I saw no hope of obtaining *a Reform of the Parliament*, without which it was clear to me, that the people of England must continue to work solely for the benefit of the great insolent families, whom I hated for their injustice and rapacity, and despised for their meanness and ignorance. I saw, in them, a mass of debauched and worthless beings, having at their command an army to compel the people to surrender to them the fruits of their industry; and, in addition, a body existing under the garb of *religion*, almost as despicable in point of character, and still more malignant.

859. I could not have died in peace, leaving my children the slaves of such a set of beings;

and, I could not live in peace, knowing that, at any hour, I might die and so leave my family. Therefore, I resolved, like the Lark in the fable, to *remove* my brood, which was still more numerous than that of the Lark. While the war was going on between England and America, I could not come to this country. Besides, I had great affairs to arrange. In 1816, having made my preparations, I set off, *not with my family;* for, that I did not think a prudent step. It was necessary for me to *see* what America really was. I, therefore, came for that purpose.

860. I was well pleased with America, over a considerable part of which I travelled. I saw an absence of human misery. I saw a government taking away a very, very small portion of men's earnings. I saw ease and happiness and a fearless utterance of thought every where prevail. I saw laws like those of the *old laws* of England, every where obeyed with cheerfulness and held in veneration. I heard of no mobs, no riots, no spies, no transportings, no hangings. I saw those very *Irish,* to keep whom in order, such murderous laws exist in Ireland, here good, peaceable, industrious citizens. I saw no placemen and pensioners, riding the people under foot. I saw no greedy Priesthood, fattening on the fruits of labour in which they had never participated, and which fruits

they seized in despite of the people. I saw a *Debt*, indeed, but then, it was so insignificant a thing; and, besides, it had been contracted for *the people's use*, and not for that of a set of tyrants, who had used the money to *the injury of the people*. In short, I saw a state of things, precisely the reverse of that in England, and very nearly what it would be in England, if the Parliament were reformed.

861. Therefore, in the Autumn of 1816, I returned to England fully intending to return the next spring with my family and whatever I possessed of the fruits of my labours, and to make America my country and the country of that family. Upon my return to England, however, I found a great stir about *Reform;* and, having, in their full force, all those feelings, which make our native country dear to us, I said, at once, "my desire "is, not to change country or countrymen, but "to change slavery for freedom: give me free- "dom here, and here I'll remain." These are nearly the very words that I uttered to Mr. COBBETT, when first introduced to him, in December, 1816, by that excellent man, MAJOR CARTWRIGHT. Nor was I unwilling to *labour myself* in the cause of Reform. I was one of those very *Delegates*, of whom the Borough-tyrants said so many falsehoods, and whom SIR FRANCIS BURDETT so shamefully abandoned.

In the meeting of Delegates, I thought we went too far in reposing confidence in him: I spoke my opinion as to this point: and, in a very few days, I had the full proof of the correctness of my opinion. I was present when MAJOR CARTWRIGHT opened a letter from SIR FRANCIS, which had come from *Leicestershire.* I thought the kind-hearted old Major would have dropped upon the floor! I shall never forget his looks as he read that letter. If the paltry Burdett had a hundred lives, the taking of them all away would not atone for the pain he that day gave to Major Cartwright, not to mention the pain given to others, and the injury done to the cause. For my part, I was not much disappointed. I had no opinion of Sir Francis Burdett's being sound. He seemed to me too much attached to his *own importance* to do the people any real service. He is an *aristocrat;* and that is enough for me. It is folly to suppose, that such a man will *ever* be a real friend of the rights of the people. I wish he were *here* a little while. He would soon find his proper level; and that would not, I think, be very high. Mr. HUNT was very much against our confiding in BURDETT; and he was perfectly right. I most sincerely hope, that my countrymen will finally destroy the tyrants who oppress them; but, I am very sure, that, before they succeed in it, they must cure themselves of the folly of depending for assistance on the *nobles* or the *half-nobles*.

862. After witnessing this conduct in Burdett, I set off home, and thought no more about effecting a Reform. The *Acts* that soon followed were, by me, looked upon as *matters of course.* The tyranny could go on no longer *under disguise.* It was compelled to shew its naked face; but, it is now, in reality, not worse than it was before. It now does no more than rob the people, and that it did before. It kills more now out-right; but, men may as well be shot, or stabbed, or hanged, as starved to death.

863. During the Spring and the early part of the Summer, of 1817, I made preparations for the departure of myself and family, and when all was ready, I bid an everlasting adieu to Boroughmongers, Sinecure placemen and place-women, pensioned Lords and Ladies, Standing Armies in time of peace, and (rejoice, oh! my children!) to a hireling, tithe-devouring Priesthood. We arrived safe and all in good health, and which health has never been impaired by the climate. We are in a state of ease, safety, plenty; and how can we help being as happy as people can be? The more I see of my adopted country, the more gratitude do I feel towards it for affording me and my numerous

offspring protection from the tyrants of my native country. There I should have been in constant anxiety about my family. Here I am in none at all. Here I am in fear of no *spies*, no *false witnesses*, no *blood-money men*. Here no fines, irons, or gallowses await me, let me *think* or *say* what I will about the government. Here I have to pay no people to be ready to shoot at me, or run me through the body, or chop me down. Here no vile Priest can rob me and mock me in the same breath.

864. In the year 1816 my travelling in America was confined to the Atlantic States. I there saw enough to determine the question of emigration or no emigration. But, a spot *to settle on myself* was another matter; for, though I do not know, that I shall meddle with any sort of trade, or occupation, in the view of getting money, I ought to look about me, and to consider soberly as to a spot to *settle on* with so large a family. It was right, therefore, for me to see the *Western Countries*. I have done this; and the particulars, which I thought worthy my notice, I noted down in a *Journal*. This Journal I now submit to the public. My chief motive in the publication is to endeavour to convey useful information, and especially to those persons, who may be disposed to follow my example, and to withdraw their families

and fortunes from beneath the hoofs of the tyrants of England.

865. I have not the vanity to suppose myself *eminently* qualified for any thing beyond my own profession; but I have been an attentive observer; I have raised a considerable fortune by my own industry and economy; I have, all my life long, studied the matters connected with agriculture, trade, and manufactures. I had a desire to acquire an accurate knowledge of the Western Countries, and what I did acquire I have endeavoured to communicate to others. It was not my object to give flowery descriptions. I leave that to poets and painters. Neither have I attempted any *general* estimate of the means or manner of living, or getting money, in the West. But, I have contented myself with merely noting down the facts that struck me; and from those facts the reader must draw his conclusions.

866. In one respect I am a proper person to give an account of the Western Countries. I have *no lands there:* I have no *interests* there: I have nothing to warp my judgment in favour of those countries: and yet, I have as little in the Atlantic States to warp my judgment in their favour. I am perfectly impartial in my feelings, and am, therefore, likely to be impartial in my words. My good wishes extend to

the utmost boundary of my adopted country. Every particular part of it is as dear to me as every other particular part.

867. I have recommended most strenuously the encouraging and promoting of *Domestic Manufacture;* not because I mean to be engaged in any such concern myself; for it is by no means likely that I ever shall; but, because I think that such encouragement and promotion would be greatly beneficial to America, and because it would provide a happy Asylum for my native oppressed and distressed countrymen, who have been employed all the days of their lives in manufactures in England, where the principal part of the immense profits of their labour is consumed by the Borough tyrants and their friends, and expended for the vile purpose of perpetuating a system of plunder and despotism at home, and all over the world.

868. Before I conclude this Introduction, I must observe, that I see with great pain, and with some degree of shame, the behaviour of some persons from England, who appear to think that they give proof of their *high breeding* by repaying civility, kindness, and hospitality, with *reproach* and *insolence.* However, these persons are *despised.* They produce very little impression here; and, though the accounts

they send to England, may be believed by some, they will have little effect on persons of sense and virtue. *Truth* will make its way; and it is, thank God, now making its way with great rapidity.

869. I could mention numerous instances of Englishmen, coming to this country with hardly a dollar in their pocket, and arriving at a state of ease and plenty and even riches in a few years; and I explicitly declare, that I have never known or heard of, an instance of one common labourer who, with common industry and economy, did not greatly better his lot. Indeed, how can it otherwise be, when the average wages of agricultural labour is *double* what it is in England, and when the average price of food is not more than half what it is in that country? These two facts, undeniable as they are, are quite sufficient to satisfy any man of sound mind.

870. As to the *manners* of the people, they are precisely to my taste: unostentatious and simple. Good sense I find every where, and never affectation. Kindness, hospitality, and never-failing civility. I have travelled more than four thousand miles about this country; and I never met with one single insolent or rude native American.

871. I trouble myself very little about the

party politics of the country. These contests are the natural offspring of freedom ; and, they tend to perpetuate that which produces them. I look at the people as a *whole ;* and I love them and feel grateful to them for having given the world a practical proof, that peace, social order, and general happiness can be secured, and best secured, without Monarchs, Dukes, Counts, Baronets, and Knights. I have no unfriendly feeling towards any Religious Society. I wish well to every member of every such Society ; but, I love the Quakers, and feel grateful towards them, for having proved to the world, that all the virtues, public as well as private, flourish most and bring forth the fairest fruits when unincumbered with those noxious weeds, hireling priests.

THOMAS HULME.

THE JOURNAL.

---

872. *PITTSBURGH*, *June* 3. — Arrived here with a friend as travelling companion, by the mail stage from Philadelphia, after a journey of six days; having set out on the 28th May. We were much pleased with the face of the country, the greatest part of which was new to me. The route, as far as Lancaster, lay through a rich and fertile country, well cultivated by good, settled proprietors; the road excellent: smooth as the smoothest in England, and hard as those made by the cruel *corvées* in France. The country finer, but the road not always so good, all the way from Lancaster, by Little York, to Chambersburgh; after which it changes for mountains and poverty, except in timber. Chambersburgh is situated on the North West side of that fine valley which lies between the South and North Mountains, and which extends from beyond the North East boundary of Pennsylvania to nearly the South West extremity of North Carolina, and which has limestone for its bottom and rich and fertile

soil, and beauty upon the face of it, from one end to the other. The ridges of mountains called the Allegany, and forming the highest land in North America between the Atlantic and Pacific oceans, begin here and extend across our route nearly 100 miles, or, rather, *three days*, for it was no less than half the journey to travel over them; they rise one above the other as we proceed Westward, till we reach the Allegany, the last and most lofty of all, from which we have a view to the West farther than the eye can carry. I can say nothing in commendation of the road over these mountains, but I must admire the drivers, and their excellent horses. The road is every thing that is bad, but the skill of the drivers, and the well constructed vehicles, and the capital old English horses, overcome every thing. We were rather singularly fortunate in not breaking down or upsetting; I certainly should not have been surprized if the whole thing, horses and all, had gone off the road and been dashed to pieces. A new road is making, however, and when that is completed, the journey will be shorter in point of time, just one half. A fine even country we get into immediately on descending the Allegany, with very little appearance of unevenness or of barrenness all the way to Pittsburgh; the evidence of good land in the crops, and the country beautified by a various mixture of woods and fields.

873. Very good accommodations for travellers the whole of the way. The stage stops to breakfast and to dine, and sleeps where it sups. They literally feasted us every where, at every meal, with venison and good meat of all sorts: every thing in profusion. In one point, however, I must make an exception, with regard to some houses: at night I was surprized, in taverns so well kept in other respects, to find bugs in the beds! I am sorry to say I observed (or, rather, *felt*,) this too often. Always good eating and drinking, but not always good sleeping.

874. *June 4th and 5th.*—Took a view of Pittsburgh. It is situated between the mouths of the rivers Allegany and Monongahela, at the point where they meet and begin the Ohio, and is laid out in a triangular form, so that two sides of it lie contiguous to the water. Called upon Mr. Bakewell, to whom we were introduced by letter, and who very obligingly satisfied our curiosity to see every thing of importance. After showing us through his extensive and well conducted glass works, he rowed us across the Monongahela to see the mines from which the fine coals we had seen burning were brought. These coals are taken out from the

side of a steep hill, very near to the river, and brought from thence and laid down in any part of the town for 7 cents the bushel, weighing, perhaps, 80 lbs. Better coals I never saw. A bridge is now building over the river, by which they will most probably be brought still cheaper.

875. This place surpasses even my expectations, both in natural resources and in extent of manufactures. Here are the materials for every species of manufacture, nearly, and of excellent quality and in profusion; and these means have been taken advantage of by skilful and industrious artizans and mechanics from all parts of the world. There is scarcely a denomination of manufacture or manual profession that is not carried on to a great extent, and, as far as I have been able to examine, in the best manner. The manufacture of iron in all the different branches, and the mills of all sorts, which I examined with the most attention, are admirable.

876. Price of flour, from 4 to 5 dollars a barrel; butter 14 cents per lb.; other provisions in proportion and mechanic's and good labourer's wages 1 dollar, and ship-builder's 1 dollar and a half, a day.

877. *June 6th.*—Leave Pittsburgh, and set out in a thing called an ark, which we buy for

the purpose, down the Ohio. We have, besides, a small skiff, to tow the ark and go ashore occasionally. This ark, which would stow away eight persons, close packed, is a thing by no means pleasant to travel in, especially at night. It is strong at bottom, but may be compared to an orange-box, bowed over at top, and so badly made as to admit a boy's hand to steal the oranges: it is proof against the river, but not against the rain.

878. Just on going to push off the wharf, an English officer stepped on board of us, with all the curiosity imaginable. I at once took him for a spy hired to way-lay travellers. He began a talk about the Western countries, anxiously assuring us that we need not hope to meet with such a thing as a respectable person, travel where we would. I told him I hoped in God I should see no spy or informer, whether in plain clothes or regimentals, and that of one thing I was certain, at any rate: that I should find no Sinecure placeman or pensioner in the Western country.

879. The Ohio, at its commencement, is about 600 yards broad, and continues running with nearly parallel sides, taking two or three different directions in its course, for about 200 miles. There is a curious contrast between the waters which form this river: that of the Alle-

gany is clear and transparent, that of the Monongahela thick and muddy, and it is not for a considerable distance that they entirely mingle. The sides of the river are beautiful; there are always rich bottom lands upon the banks, which are steep and pretty high, varying in width from a few yards to a mile, and skirted with steep hills varying also in height, overhanging with fine timber.

880. *June 7th.*—Floating down the Ohio, at the rate of four miles an hour. Lightning, thunder, rain and hail pelting in upon us. The hail-stones as large as English hazle-nuts. Stop at Steubenville all night. A nice place; has more stores than taverns, which is a good sign.

881. *June 8th.*—Came to Wheeling at about 12 o'clock. It is a handsome place, and of considerable note. Stopped about an hour. Found flour to be about 4 to 5 dollars a barrel; fresh beef 4 to 6 cents per lb., and other things (the produce of the country) about the same proportion. Labourers' wages, 1 dollar a day. Fine coals here, and at Steubenville.

882. *June 9th.*—Two fine young men join us, one a carpenter and the other a saddler, from Washington, in a skiff that they have bought at Pittsburgh, and in which they are taking a journey of about 700 miles down the river. We allow them to tie their skiff to our ark, for

which they very cheerfully assist us. Much diverted to see the nimbleness with which they go on shore sometimes with their rifles to shoot pigeons and squirrels. The whole expences of these two young men in floating the 700 miles, will be but 7 dollars each, including skiff and every thing else.

883. This day pass Marietta, a good looking town at the mouth of the Muskingham River. It is, however, like many other towns on the Ohio, built on too low ground, and is subject to inundations. Here I observe a contrivance of great ingenuity. There is a strong rope put across the mouth of the river, opposite the town, fastened to trees or large posts on each side; upon this rope runs a pulley or block, to which is attached a rope, and to the rope a ferry-boat, which, by moving the helm first one way and then the other, is propelled by the force of the water across the river backwards or forwards.

884. *June* 10*th.*—Pass several fine coal mines, which, like those at Pittsburgh, Steubenville, Wheeling and other places, are not above 50 yards from the river and are upwards of 10 yards above high water. The river now becomes more winding than we have hitherto found it. It is sometimes so serpentine that it appears before and behind like a continuation of lakes, and the

hills on its banks seem to be the separations. Altogether, nothing can be more beautiful.

885. *June* 11*th.*—A very hot day, but I could not discover the degree of heat. On going along we bought two Perch, weighing about 8 lbs. each, for 25 cents, of a boy who was fishing. Fish of this sort will sometimes weigh 30 lbs. each.

886. *June* 12*th.*—Pass Portsmouth, at the mouth of the Scioto River. A sort of village, containing a hundred or two of houses. Not worthy of any particular remark.

887. *June* 13*th.*—Arrived at Cincinnati about midnight. Tied our ark to a large log at the side of the river, and went to sleep. Before morning, however, the fastening broke, and, if it had not been for a watchful back-woods-man whom we had taken on board some distance up the river, we might have floated ten or fifteen miles without knowing it. This back-woods-man, besides being of much service to us, has been a very entertaining companion. He says he has been in this country forty years, but that he is an Englishman, and was bred in Sherwood Forest (he could not have come from a better nursery). All his adventures he detailed to us very minutely, but dwelt with particular warmth upon one he had had with a priest, lately, who, to spite him for preaching, brought

an action against him, but was cast and had to pay costs.

888. *June 14th and 15th.*—Called upon Doctor Drake and upon a Mr. Bosson, to whom we had letters. These gentlemen shewed us the greatest civility, and treated us with a sort of kindness which must have changed the opinion even of the English officer whom we saw at Pittsburgh, had he been with us. I could tell that dirty hireling scout, that even in this short space of time, I have had the pleasure to meet many gentlemen, very well informed, and possessing great knowledge as to their own country, evincing public spirit in all their actions, and hospitality and kindness in all their demeanor; but, if they be pensioners, male or female, or sinecure place lords or ladies, I have yet come across, thank God, no *respectable people.*

889. Cincinnati is a very fine town, and elegantly (not only in the American acceptation of the word) situated on the banks of the river, nearly opposite to Licking Creek, which runs out of Kentucky, and is a stream of considerable importance. The country round the town is beautiful, and the soil rich; the fields in its immediate vicinity bear principally grass, and clover of different sorts, the fragrant smell of which perfumes the air. The town itself ranks

next to Pittsburgh, of the towns on the Ohio, in point of manufactures.

890. We sold our ark, and its produce formed a deduction from our expences, which, with that deduction, amounted to 14 dollars each, including every thing, for the journey from Pittsburgh to this place, which is upwards of 500 miles. I could not but remark the price of fuel here; 2 dollars a cord for Hickory; a cord is 8 feet by 4, and 4 deep, and the wood, the best in the world; it burns much like green Ash, but gives more heat. This, which is of course the highest price for fuel in this part of the country, is only about a fifth of what it is at Philadelphia.

891. *June 16th.*—Left Cincinnati for Louisville with seven other persons, in a skiff about 20 feet long and 5 feet wide.

892. *June 17th.*—Stopped at VEVAY, a very neat and beautiful place, about 70 miles above the falls of the Ohio. Our visit here was principally to see the mode used, as well as what progress was made, in the cultivation of the vine, and I had a double curiosity, never having as yet seen a vineyard. These vineyards are cultivated entirely by a small settlement of Swiss, of about a dozen families, who have been here about ten years. They first settled on the Kentucky river, but did not succeed

there. They plant the vines in rows, attached to stakes like espaliers, and they plough between with a one-horse plough. The grapes, which are of the sorts of claret and madeira, look very fine and luxuriant, and will be ripe in about the middle of September. The soil and climate both appear to be quite congenial to the growth of the vine: the former rich and the latter warm. The north west wind, when it blows, is very cold, but the south, south east and south west winds, which are always warm, are prevalent. The heat, in the middle of the summer, I understand, is very great, being generally above 85 degrees, and sometimes above 100 degrees. Each of these families has a farm as well as a vineyard, so that they supply themselves with almost every necessary and have their wine all clear profit. Their produce will this year be probably not less than 5000 gallons; we bought 2 gallons of it at a dollar each, as good as I would wish to drink. Thus it is that the tyrants of Europe create vineyards in this new country!

893. *June 18th.*—Arrived at Lousville, Kentucky. The town is situated at the commencement of the falls, or rapids, of the Ohio. The river, at this place, is little less than a mile wide, and the falls continue from a ledge of rocks which runs across the river in a sloping

direction at this part, to Shippingport, about 2 miles lower down. Perceiving stagnant waters about the town, and an appearance of the house that we stopped at being infested with bugs, we resolved not to make any stay at Louisville, but got into our skiff and floated down the falls to Shippingport. We found it very rough floating, not to say dangerous. The river of very unequal widths and full of islands and rocks along this short distance, and the current very rapid, though the descent is not more than 22 feet. At certain times of the year the water rises so that there is no fall; large boats can then pass.

894. At Shippingport, stopped at the house of Mr. Berthoud, a very respectable French gentleman, from whom we received the greatest civility during our stay, which was two nights and the day intervening.

895. Shippingport is situated at a place of very great importance, being the upper extremity of that part of the river which is navigable for heavy steam-boats. All the goods coming from the country are re-shipped, and every thing going to it is un-shipped, here. Mr. Berthoud has the store in which the articles exporting or importing are lodged; and is, indeed, a great shipper, though at a thousand miles from the sea.

896. *June 20th.*—Left the good and comfortable house of Mr. Berthoud, very much pleased with him and his amiable wife and family, though I differed with him a little in politics. Having been taught at church, when a boy, that the Pope was the whore of Babylon, that the Bourbons were tyrants, and that the Priests and privileged orders of France were impostors and petty tyrants under them, I could not agree with him in applauding the Boroughmongers of England for re-subjugating the people of France, and restoring the Bourbons, the Pope, and the Inquisition.

897. Stop at New Albany, 2 miles below Shippingport, till the evening. A Mr. Paxton, I am told, is the proprietor of a great part of the town, and has the grist and saw-mills, which are worked by steam, and the ferry across the river. Leave this place in company with a couple of young men from the western part of the state of New York, who are on their way to Tennessee in a small ferry-boat. Their whole journey will, probably, be about 1,500 miles.

898. *June 21st.*—Floating down the river, without any thing in particular occurring.

899. *June 22nd.*—Saw a Mr. Johnstone and his wife reaping wheat on the side of the river. They told us they had come to this spot last

year, direct from Manchester, Old England, and had bought their little farm of 55 acres of a back-woods-man who had cleared it, and was glad to move further westward, for 3 dollars an acre. They had a fine flock of little children, and pigs and poultry, and were cheerful and happy, being confident that their industry and economy would not be frustrated by visits for tithes or taxes.

900. *June* 23*rd.*—See great quantities of turkey-buzzards and thousands of pigeons. Came to Pigeon Creek, about 230 miles below the Falls, and stopped for the night at Evansville, a town of nine months old, near the mouth of it. We are now frequently met and passed by large, fine steam-boats, plying up and down the river. One went by us as we arrived here which had left Shippingport only the evening before. They go down the river at the rate of 10 miles an hour, and charge passengers 6 cents a mile, boarding and lodging included. The price is great, but the time is short.

901. *June* 24*th.*—Left Evansville. This little place is rapidly increasing, and promises to be a town of considerable trade. It is situated at a spot which seems likely to become a port for shipping to Princeton and a pretty large district of Indiana. I find that the land speculators have made entry of the most eligible

tracts of land, which will impede the partial, though not the final, progress of population and improvement in this part of the state.

902. On our way to Princeton, we see large flocks of fine wild turkeys, and whole herds of pigs, apparently very fat. The pigs are wild also, but have become so from neglect. Some of the inhabitants, who prefer sport to work, live by shooting these wild turkeys and pigs, and, indeed, sometimes, I understand, they shoot and carry off those of their neighbours before they are wild.

903. *June 25th.*—Arrived at Princeton, Indiana, about 20 miles from the river. I was sorry to see very little doing in this town. They cannot *all* keep stores and taverns! One of the store-keepers told me he does not sell more than ten thousand dollars value per annum: he ought, then, to manufacture something and not spend nine tenths of his time in lolling with a segar in his mouth.

904. *June 26th.*—At Princeton, endeavouring to purchase horses, as we had now gone far enough down the Ohio. While waiting in our tavern, two men called in armed with rifles, and made enquiries for some horses they suspected to be stolen. They told us they had been almost all the way from Albany, to Shawnee town after them, a distance of about 150

miles. I asked them how they would be able to secure the thieves, if they overtook them, in these wild woods; "O" said they, "shoot "them off the horses." This is a summary mode of executing justice, thought I, though probably the most effectual, and, indeed, only one in this state of society. A thief very rarely escapes here; not nearly so often as in more populous districts. The fact was, in this case, however, we discovered afterwards, that the horses had strayed away, and had returned home by this time. But, if they had been stolen, the stealers would not have escaped. When the loser is tired, another will take up the pursuit, and the whole country is up in arms till he is found.

905. *June 27th.*—Still at Princeton. At last we get suited with horses. Mine costs me only 135 dollars with the bridle and saddle, and that I am told is 18 dollars too much.

906. *June 28th.*—Left Princeton, and set out to see Mr. Birkbeck's settlement, in Illinois, about 35 miles from Princeton. Before we got to the Wabash we had to cross a swamp of half a mile wide; we were obliged to lead our horses, and walk up to the knees in mud and water. Before we got half across we began to think of going back; but, there is a sound bottom under it all, and we waded through it as well as

we could. It is, in fact, nothing but a bed of very soft and rich land, and only wants draining to be made productive. We soon after came to the banks of the great Wabash, which is here about half a mile broad, and as the ferry-boat was crossing over with us I amused myself by washing my dirty boots. Before we mounted again we happened to meet with a neighbour of Mr. Birkbeck's, who was returning home; we accompanied him, and soon entered into the prairie lands, up to our horses' bellies in fine grass. These prairies, which are surrounded with lofty woods, put me in mind of immense noblemen's parks in England. Some of those we passed over are called *wet prairies*, but, they are dry at this time of the year; and, as they are none of them flat, they need but very simple draining to carry off the water all the year round. Our horses were very much tormented with flies, some as large as the English horse-fly and some as large as the wasp; these flies infest the prairies that are unimproved about three months in the year, but go away altogether as soon as cultivation begins.

907. Mr. Birkbeck's settlement is situated between the two Wabashes, and is about ten miles from the nearest navigable water; we arrived there about sun-set, and met with a welcome which amply repaid us for our day's toil.

We found that gentleman with his two sons perfectly healthy and in high spirits: his daughters were at Henderson (a town in Kentucky, on the Ohio) on a visit. At present his habitation is a cabin, the building of which cost only 20 dollars; this little hutch is near the spot where he is about to build his house, which he intends to have in the most eligible situation in the prairie for convenience to fuel and for shelter in winter, as well as for breezes in summer, and will, when that is completed, make one of its appurtenances. I like this plan of keeping the old log-house; it reminds the grand children and their children's children of what their ancestor has done for their sake.

908. Few settlers had as yet joined Mr. Birkbeck; that is to say, settlers likely to become "*society*;" he has labourers enough near him, either in his own houses or on land of their own joining his estate. He was in daily expectation of his friends Mr. Flower's family, however, with a large party besides; they had just landed at Shawnee Town, about 20 miles distant. Mr. Birkbeck informs me he has made entry of a large tract of land, lying, part of it, all the way from his residence to the great Wabash; this he will re-sell again in lots to any of his friends, they taking as much of it and wherever they choose (provided it be no more than they can

cultivate), at an advance which I think very fair and liberal.

909. The whole of his operations had been directed hitherto (and wisely in my opinion) to building, fencing, and other important preparations. He had done nothing in the cultivating way but make a good garden, which supplies him with the only things that he cannot purchase, and, at present, perhaps, with more economy than he could grow them. He is within twenty miles of Harmony, in Indiana, where he gets his flour and all other necessaries (the produce of the country), and therefore employs himself much better in making barns and houses and mills for the reception and disposal of his crops, and fences to preserve them while growing, *before he grows them*, than to *get the crops first*. I have heard it observed that *any* American settler, even without a dollar in his pocket, would have *had something growing by this time*. Very true! I do not question that at all; for, the very first care of a settler without a dollar in his pocket is to get something to eat, and, he would consequently set to work scratching up the earth, fully confident that after a long summering upon wild flesh (without salt, perhaps) his own belly would stand him for barn, if his jaws would not for mill. But the case is very different with Mr. Birkbeck, and at present he

has need for no other provision for winter but about a three hundredth part of his fine grass turned into hay, which will keep his necessary horses and cows; besides which he has nothing that eats but such pigs as live upon the waste, and a couple of fine young deer (which would weigh, they say when full grown, 200 lbs. dead weight), that his youngest son is rearing up as pets.

910. I very much admire Mr. Birkbeck's mode of *fencing*. He makes a ditch 4 feet wide at top, sloping to 1 foot wide at bottom, and 4 feet deep. With the earth that comes out of the ditch he makes a bank on one side, which is turfed towards the ditch. Then a long pole is put up from the bottom of the ditch to 2 feet above the bank; this is crossed by a short pole from the other side, and then a rail is laid along between the forks. The banks were growing beautifully, and looked altogether very neat as well as formidable; though a live hedge (which he intends to have) instead of dead poles and rails, upon top, would make the fence far more effectual as well as handsomer. I am always surprized, until I reflect how universally and to what a degree, farming is neglected in this country, that this mode of fencing is not adopted in cultivated districts, especially where the land is wet, or lies low; for, there it answers

a double purpose, being as effectual a drain as it is a fence.

911. I was rather disappointed, or sorry, at any rate, not to find near Mr. Birkbeck's any of the means for machinery or of the materials for manufactures, such as the water-falls, and the minerals and mines, which are possessed in such abundance by the states of Ohio and Kentucky, and by some parts of Pennsylvania. Some of these, however, he may yet find. Good water he has, at any rate. He showed me a well 25 feet deep, bored partly through hard substances near the bottom, that was nearly overflowing with water of excellent quality.

912. *July* 1*st*.—Left Mr. Birkbeck's for Harmony, Indiana. The distance by the direct way is about 18 miles, but there is no road, as yet: indeed, it was often with much difficulty that we could discover the way at all. After we had crossed the Wabash, which we did at a place called Davis's Ferry, we hired a man to conduct us some part of the way through the woods. In about a mile he brought us to a track, which was marked out by slips of bark being stripped off the trees, once in about 40 yards; he then left us, and told us we could not mistake if we followed that track. We soon lost all appearance of the track, however, and of the "*blazing*" of the trees, as they call it; but, as

it was useless to go back again for another guide, our only way was to keep straight on in the same direction, bring us where it would. Having no compass, this nearly cost us our sight, for it was just mid-day, and we had to gaze at the sun a long time before we discovered what was our course. After this we soon, to our great joy, found ourselves in a large corn field; rode round it, and came to Johnson's Ferry, a place where a Bayou *(Boyau)* of the Wabash is crossed. This Bayou is a run out of the main river, round a flat portion of land, which is sometimes overflowed: it is part of the same river, and the land encompassed by it, an island. Crossed this ferry in a canoe, and got a ferry-man to swim our horses after us. Mounted again and followed a track which brought us to Black River, which we forded without getting wet, by holding our feet up. After crossing the river we found a man who was kind enough to shew us about half a mile through the woods, by which our journey was shortened five or six miles. He put us into a direct track to Harmony, through lands as rich as a dung-hill, and covered with immense timber; we thanked him, and pushed on our horses with eager curiosity to see this far-famed Harmonist Society.

913. On coming within the precincts of the Harmonites we found ourselves at the side of the Wabash again; the river on our right hand, and their lands on our left. Our road now lay across a field of Indian corn, of, at the very least, a mile in width, and bordering the town on the side we entered; I wanted nothing more than to behold this immense field of most beautiful corn to be at once convinced of all I had heard of the industry of this society of Germans, and I found, on proceeding a little farther, that the progress they had made exceeded all my idea of it.

914. The town is methodically laid out in a situation well chosen in all respects; the houses are good and clean, and have, each one, a nice garden well stocked with all vegetables and tastily ornamented with flowers. I observe that these people are very fond of flowers, by the bye; the cultivation of them, and musick, are their chief amusements. I am sorry to see this, as it is to me a strong symptom of simplicity and ignorance, if not a badge of their German slavery. Perhaps the pains they take with them is the cause of their flowers being finer than any I have hitherto seen in America, but, most probably, the climate here is more favourable. Having refreshed ourselves at the Tavern, where we

found every thing we wanted for ourselves and our horses, and all very clean and nice, besides many good things we did not expect, such as beer, porter, and even wine, all made within the Society, and very good indeed, we then went out to see the people at their harvest, which was just begun. There were 150 men and women all reaping in the same field of wheat. A beautiful sight! The crop was very fine, and the field, extending to about two miles in length, and from half a mile to a mile in width, was all open to one view, the sun shining on it from the West, and the reapers advancing regularly over it.

915. At sun-set all the people came in, from the fields, work-shops, mills, manufactories, and from all their labours. This being their evening for prayer during the week, the Church bell called them out again, in about 15 minutes, to attend a lecture from their High Priest and Law-giver, Mr. George Rapp. We went to hear the lecture, or, rather, to see the performance, for, it being all performed in German, we could understand not a word. The people were all collected in a twinkling, the men at one end of the Church and the women at the other; it looked something like a Quaker Meeting, except that there was not a single little child in the place. Here they were kept by

their Pastor a couple of hours, after which they returned home to bed. This is the quantum of Church-service they perform during the week; but on Sundays they are in Church nearly the whole of the time from getting up to going to bed. When it happens that Mr. Rapp cannot attend, either by indisposition or other accident, the Society still meet as usual, and the *elders* (certain of the most trusty and discreet, whom the Pastor selects as a sort of assistants in his divine commission) converse on religious subjects.

916. Return to the Tavern to sleep; a good comfortable house, well kept by decent people, and the master himself, who is very intelligent and obliging, is one of the very few at Harmony who can speak English. Our beds were as good as those stretched upon by the most highly pensioned and placed Boroughmongers, and our sleep, I hope, much better than the tyrants ever get, in spite of all their dungeons and gags.

917. *July 2nd.*—Early in the morning, took a look at the manufacturing establishment, accompanied by our Tavern-keeper. I find great attention is paid to this branch of their affairs. Their principle is, not to be content with the profit upon the manual labour of *raising* the article, but also to have the benefit of the ma-

chine in preparing it for *use.* I agree with them perfectly, and only wish the subject was as well understood all over the United States as it is at Harmony. It is to their skill in this way that they owe their great prosperity; if they had been nothing but farmers, they would be now at Harmony in Pennsylvania, poor cultivators, getting a bare subsistence, instead of having doubled their property two or three times over, by which they have been able to move here and select one of the choicest spots in the country.

918. But, in noting down the state of this Society, as it now is, its *origin* should not be forgotten; the curious history of it serves as an explanation to the jumble of sense and absurdity in the association. I will therefore trace the Harmonist Society from its outset in Germany to this place.

919. The Sect had its origin at Wurtemberg in Germany, about 40 years ago, in the person of its present Pastor and Master, George Rapp, who, by his own account, "having long seen "and felt the decline of the Church, found "himself impelled to bear testimony to the "fundamental principles of the Christian Reli- "gion; and, finding no toleration for his in- "spired doctrines, or for those who adopted "them, he determined with his followers to go

" to that part of the earth where they were
" free to worship God according to the dictates
" of their conscience." In other words (I suppose), he had long beheld and experienced the slavery and misery of his country, and, feeling in his conscience that he was born more for a ruler than for a slave, found himself imperiously called upon to collect together a body of his poor countrymen and to lead them into a land of liberty and abundance. However, allowing him to have had no other than his professed views, he, after he had got a considerable number of proselytes, amounting to seven or eight hundred persons, among whom were a sufficiency of good labourers and artizans in all the essential branches of workmanship and trade, besides farmers, he embodied them into a Society, and then came himself to America (not trusting to Providence to lead the way) to seek out the land destined for these chosen children. Having done so, and laid the plan for his route to the land of peace and Christian love, with a foresight which shows him to have been by no means unmindful to the *temporal* prosperity of the Society, he then landed his followers in separate bodies, and prudently led them in that order to a resting place within Pennsylvania, choosing rather to retard their progress through the wilderness than to hazard

the discontent that might arise from want and fatigue in traversing it at once. When they were all arrived, Rapp constituted them into one body, having every thing in common, and called the settlement *Harmony*. This constitution he found authorized by the passage in Acts, iv, 32. "And the multitude of them that "believed were of one heart, and of one soul: "neither said any of them that aught of the "things he possessed was his own, *but that* "*they had all things common.*" Being thus associated, the Society went to work, early in 1805, building houses and clearing lands, according to the order and regulations of their leader; but, the community of stock, or the regular discipline, or the restraints which he had reduced them to, and which were essential to his project, soon began to thin his followers, and principally, too, those of them who had brought most substance into the society; they demanded back their original portions and set out to seek the Lord by themselves. This falling off of the society, though it was but small, comparatively, in point of numbers, was a great reduction from their means; they had calculated what they should want to consume, and had laid the rest out in land; so that the remaining part were subjected to great hardships and difficulties for the first year or two of their settling, which was during the time of their greatest labours. However, it was not long before they began to reap the fruits of their toil, and in the space of six or seven years their settlement became a most flourishing colony. During that short space of time they brought into cultivation 3,000 acres of land (a third of their whole estate), reared a flock of nearly 2,000 sheep, and planted hop-gardens, orchards, and vineyards; built barns and stables to house their crops and their live stock, granaries to keep one year's produce of grain always in advance, houses to make their cyder, beer, and wine in, and good brick or stone warehouses for their several species of goods; constructed distilleries, mills for grinding, sawing, making oil, and, indeed, for every purpose, and machines for manufacturing their various materials for clothing and other uses; they had, besides, a store for retailing Philadelphia goods to the country, and nearly 100 good dwelling-houses of wood, a large stone-built tavern, and, as a proof of superabundance, a dwelling-house and a meeting-house (alias the parsonage and church) which they had neatly built of brick. And; besides all these improvements within the society, they did a great deal of business, principally in the way of manufacturing, for the people of the country. They worked for them with their mills and machines, some of which did nothing else, and their blacksmiths, tailors, shoe-makers, &c. when not employed by themselves, were constantly at work for their neighbours. Thus this everlastingly-at-work band of emigrants increased their stock before they quitted their first colony, to upwards of two hundred thousand dollars, from, probably, not one fifth of that sum. What will not unceasing perseverance accomplish? But, with judgment and order to direct it, what in the world can stand against it!*

920. In comparing the state of this society as it now is with what it was in Pennsylvania, it is just the same as to *plan;* the temporal and spiritual affairs are managed in the same way, and upon the same principles, only both are more flourishing. Rapp has here brought his disciples into richer land, and into a situation better in every respect, both for carrying on their trade, and for keeping to their faith; their vast extent of land is, they say, four feet deep of rich mould, nearly the whole of it, and it lies along the banks of a fine navigable river on one side, while the possibility of much interruption from other classes of Christians is effectually guarded against by an endless barricado of woods on the other side. Bringing the means

* A more detailed account of this society, up to the year 1811, will be found in Mr. Mellish's Travels, vol 2.

and experience acquired at their first establishment, they have of course gone on improving and increasing (not in *population)* at a much greater rate. One of their greatest improvements, they tell me, is the working of their mills and manufacturing machines by steam; they feel the advantage of this more and more every year. They are now preparing to build a steam-boat; this is to be employed in their traffick with New Orleans, carrying their own surplus produce and returning with tea, coffee, and other commodities for their own consumption, and to retail to the people of the country. I believe they advance, too, in the way of ornaments and superfluities, for the dwelling-house they have now built their pastor, more resembles a Bishop's Palace than what I should figure to myself as the humble abode of a teacher of the "fundamental principles of the Christian "Religion."

921. The government of this society is by bands, each consisting of a distinct trade or calling. They have a foreman to each band, who rules it under the general direction of the society, the law-giving power of which is in the High Priest. He cannot, however make laws without the consent of the parties. The manufacturing establishment, and the mercantile affairs and public accounts are all managed by

one person; he, I believe, is one of the sons of Rapp. They have a bank, where a separate account is kept for each person; if any one puts in money, or has put in money, he may, on certain conditions as to time, take it out again. They labour and possess in common; that is to say, except where it is not practicable or is immaterial, as with their houses, gardens, cows and poultry, which they have to themselves, each family. They also retain what property each may bring on joining the concern, and he may demand it in case of leaving the society, but *without interest.*

922. Here is certainly a wonderful example of the effects of skill, industry, and force combined: this congregation of far-seeing, ingenious, crafty, and bold, and of ignorant, simple, superstitious, and obedient, Germans, has shown what may be done. But, their example, I believe, will generally only tend to confirm this free people in their suspicion that labour is concomitant to slavery or ignorance. Instead of their improvements, and their success and prosperity altogether, producing admiration, if not envy, they have a social discipline, the thought of which reduces these feelings to ridicule and contempt: that is to say, with regard to the *mass;* with respect to their leaders, one's feelings are apt to be stronger. A fundamental

of their religious creed (" *restraining clause*," a Chancery Lawyer would call it) requires restrictions on the propagation of the species; it orders such regulations as are necessary to prevent children coming but once in a certain number of years; and this matter is so arranged that, when they come, they come in little flocks, all within the same month, perhaps, like a farmer's lambs. The Law-giver here made a famously " restraining sta-" tute" upon the law of nature! This way of expounding law seems to be a main point of his policy; he by this means keeps his associates from increasing to an unruly number within, while more are sure not to come in from without; and, I really am afraid he will go a good way towards securing a monopoly of many great improvements in agriculture, both as to principle and method. People see the fine fields of the Harmonites, but, the prospect comes damped with the idea of bondage and celibacy. It is a curious society: was ever one heard of before that did not wish to increase! This smells strong of policy; some distinct view in the leaders, no doubt. Who would be surprized if we were to see a still more curious society by and bye? A *Society Sole!* very far from improbable, if the sons of Rapp (for he has children, nevertheless, as well as Parson

Malthus) and the *Elders* were to die, it not being likely that they will renounce or forfeit their right to the common stock. We should then have societies as well as corporations vested in one person! That would be quite a novel kind of benefice! but, not the less fat. I question whether the *associated* person of Mr. Rapp would not be in possession of as fine a domain and as many good things as the *incorporated* person of an Archbishop: nay, he would rival the Pope! But, to my journal.

923. Arrive at Princeton in the evening; a good part of our road lay over the fine lands of the Harmonites. I understand, by the bye, that the title deeds to these lands are taken in the name of *Rapp and of his associates.* Poor associates: if they do but rebel! Find the same store-keepers and tavern-keepers in the same attitudes that we left them in the other day. Their legs *only a little* higher than their heads, and segars in their mouths; a fine position for business! It puts my friend in mind of the Roman posture in dining.

924. *July 3rd.*—At Princeton all day. This is a pretty considerable place; very good as to buildings; but, is too much inland to be a town of any consequence until the inhabitants do that at home which they employ merchants and foreign manufactures to do for them. Pay

1 dollar for a set of old shoes to my horse, half the price of new ones.

925. *July 4th.*—Leave Princeton; in the evening, reach a place very appropriately called Mud-holes, after riding 46 miles over lands in general very good but very little cultivated, and that little very badly; the latter part of the journey in company with a Mr. Jones from Kentucky. Nature is the agriculturist here; speculation, instead of cultivation, is the order of the day amongst men. We feel the ill effects of this in the difficulty of getting oats for our horses. However, the evil is unavoidable, if it really can be called an evil. As well might I grumble that farmers have not taken possession as complain that men of capital have. Labour is the thing wanted, but, to have that, money must come first. This Mud-holes was a sort of fort, not 4 years ago, for guarding against the Indians, who then committed great depredations, killing whole families often, men, women and children. How changeable are the affairs of this world! I have not met with a single Indian in the whole course of my route.

926. *July 5th.*—Come to Judge Chambers's, a good tavern; 35 miles. On our way, pass French Lick, a strong spring of water impregnated with salt and sulphur, and called *Lick* from its being resorted to by cattle for the salt;

close by this spring is another still larger, of fine clear lime-stone water, running fast enough to turn a mill. Some of the trees near the Judge's exhibit a curious spectacle; a large piece of wood appears totally dead, all the leaves brown and the branches broken, from being roosted upon lately by an enormous multitude of pigeons. A novel sight for us, unaccustomed to the abundance of the back-woods! No tavern but this, nor house of any description, within many miles.

927. *July 6th.*—Leave the Judge's, still in company with Mr. Jones. Ride 25 miles to breakfast, not sooner finding feed for our horses; this was at the dirty log-house of Mr. ——— who has a large farm with a grist-mill on it, and keeps his yard and stables ancle deep in mud and water. If this were not one of the healthiest climates in the world, he and his family must have died in all this filth. About 13 miles further, come to New Albany, where we stop at Mr Jenkins's, the best tavern we have found in Indiana, that at Harmony excepted.

928. *July 7th.*—Resting at New Albany. We were amused by hearing a Quaker-lady preach to the natives. Her first words were "*all the nations of the earth are of one blood.*" "So," said I to myself, "this question, which

" has so long perplexed philosophers, divines " and physicians, is now set at rest!" She proceeded to vent her rage with great vehemence against hireling priests and the trade of preaching in general, and closed with dealing out large portions of brimstone to the drunkard and still larger and hotter to those who give the bottle to drink. This part of her discourse pleased me very much, and may be a saving to me into the bargain; for, the dread of everlasting roasting added to my love of economy will (I think) prevent me making my friends tipsy. A very efficacious sermon!

929. *July* 8*th.*—Jenkins's is a good tavern, but it entertains at a high price. Our bill was 6 dollars each for a day and two nights; a shameful charge. Leave New Albany, cross the Ohio, and pass through Louisville in Kentucky again, on our way to Lexington, the capital. Stop for the night at Mr. Netherton's, a good tavern. The land hitherto is good, and the country altogether healthy, if I may judge from the people, who appear more cheerful and happy than in Indiana, always excepting Harmony. Our landlord is the picture of health and strength: 6 feet 4 inches high, weighs 300 lbs., and not fat.

930. *July* 9*th.*—Dine at Mr. Overton's tavern, on our way to Frankfort; pay half a

dollar each for an excellent dinner, with as much brandy and butter-milk as we chose to drink, and good feed for our horses. In the afternoon we have the pleasure to be overtaken by two ladies on horse-back, and have their agreeable company for a mile or two. On their turning off from our road we were very reluctantly obliged to refuse an obliging invitation to drink tea at their house, and myself the more so, as one of the ladies informed me she had married a Mr. Constantine, a gentleman from my own native town of Bolton, in Lancashire. But, we had yet so far to go, and it was getting dark. This most healthful mode of travelling is universal in the Western States, and it gives me great pleasure to see it; though, perhaps, I have to thank the badness of the roads as the cause. Arrive at Frankfort, apparently a thriving town, on the side of the rough Kentucky river. The houses are built chiefly of brick, and the streets, I understand, paved with limestone. Limestone abounds in this state, and yet the roads are not good, though better than in Indiana and Ohio, for, there, there are none. I wonder the governments of these states do not set about making good roads and bridges, and even canals. I pledge myself to be able to shew them how the money might be raised, and, moreover, to prove that the expence would be paid over and over again in almost no time. Such improvements would be income to the governments instead of expence, besides being such an incalculable benefit to the states. But, at any rate, why not *roads*, and in *this* state, too, which is so remarkable for its quality of having good road materials and rich land together, generally, all over it?

931. *July* 10*th*.—Leave Frankfort, and come through a district of fine land, very well watered, to Lexington; stop at Mr. Keen's tavern. Had the good fortune to meet Mr. Clay, who carried us to his house, about a mile in the country. It is a beautiful residence, situated near the centre of a very fine farm, which is just cleared and is coming into excellent cultivation. I approve of Mr. Clay's method very much, especially in laying down pasture. He clears away all the brush or underwood, leaving timber enough to afford a sufficiency of shade to the grass, which does not thrive here exposed to the sun, as in England and other such climates. By this means he has as fine grass and clover as can possibly grow. I could not but admire to see this gentleman, possessing so much knowledge and of so much weight in his country's affairs, so attentively promoting her not less important though more silent interests

by improving her agriculture. What pleased me still more, however, because I less expected it, was, to hear Mrs. Clay, in priding herself on the state of society, and the rising prosperity of the country, citing as a proof the decency and affluence of the trades-people and mechanics at Lexington, many of whom ride about in their own carriages. What a contrast, both in sense and in sentiment, between this lady and the wives of Legislators (as they are called), in the land of the Boroughmongers! God grant that no privileged batch ever rise up in America, for then down come the mechanics, are harnessed themselves, and half ridden to death.

932. *July* 11*th.*—This is the hottest day we have had yet. Thermometer at 90 degrees, in shade. Met a Mr. Whittemore, from Boston, loud in the praise of this climate. He informed me he had lately lost his wife and five children near Boston, and that he should have lost his only remaining child, too, a son now stout and healthy, had he not resolved instantly to try the air of the west. He is confident that if he had taken this step in time he might have saved the lives of all his family. This might be, however, and yet this climate not better than that of Boston. Spent the evening with Colonel Morrison, one of the first settlers in this state; a fine looking old gentleman, with colour in

his face equal to a London Alderman. The people here are pretty generally like that portion of the people of England who get porridge enough to eat; stout, fat, and ruddy.

933. *July 12th.*—Hotter than yesterday; thermometer at 91 degrees.

934. *July 13th.*—Leave Lexington; stop at Paris, 22 miles. A fine country all the way; good soil, plenty of limestone and no musquitoes. Paris is a healthy town, with a good deal of stir; woollen and cotton manufactures are carried on here, but upon a small scale. They are not near enough to good coal mines to do much in that way. What they do, however, is well paid for. A spinner told me he gets 83 cents per lb. for his twist, which is 33 cents more than it would fetch at New York. Stop at Mr. Timberlake's, a good house. The bar-keeper, who comes from England tells me that he sailed to Canada, but he is glad he had the means to leave Canada and come to Kentucky; he has 300 dollars a year, and board and lodging. Made enquiry after young Watson, but find he has left this place and is gone to Lexington.

935. The following is a list of the wages and prices of the most essential branches of workmanship and articles of consumption, as they are here at present.

	Dolls.	Cents.	Dolls.	Cents.
Journeymen saddlers' price for drawing on men's saddles	*1	25 to	2	50
Journeymen blacksmiths, per day	1	. –	1	25
per month	25	. –	30	
Journeymen hatters (*casters*)	1	25		
Ditto, *rorum*	1	.		
Ditto for finishing, per month and found	30	.		
Journeymen shoe-makers (*coarse*)	.	75		
Ditto, *fine*	1	25		
Ditto, for boots . . .	3	25		
Journeymen tailors, by the coat	5	.		
Stone-masons or bricklayers, per day	1	. –	1	50
Carpenters, per day, and found	1	.		
Salary for a clerk, per annum	200	. –	500	
Beef, per 100 lbs. . . .	6	.		
Flour, per barrel	6	.		

936. *July* 14*th.*—Hot again; 90 degrees.

* Or, 5*s* 7½*d.* to 11*s* 3*d. sterling.* At the present rate of exchange, a *dollar* is equivalent to 4*s*. 6*d*. sterling, and a *cent* is the hundredth part of a dollar.

Arrive at Blue Licks, close by the fine Licking Creek, 22 miles from Paris. Here is a sulphur and salt spring like that at French Lick in Indiana, which makes this a place of great resort in summer for the fashionable swallowers of mineral waters; the three or four taverns are at this time completely crowded. Salt was made till latterly at this spring, by an old Scotsman; he now attends the ferry across the Creek. Not much to be said for the country round here; it is stony and barren, what I have not seen before in Kentucky.

937. *July* 15*th.*—To Maysville, or Limestone, 24 miles. This is a place on the banks of the Ohio, and is a sort of port for shipping *down* the river to a great part of that district of the state for which Louisville is the shipping port to and from New Orleans. Still hot; 90 degrees again. This is the fifth day; rather unusual, this continuance of heat. The hot spells as well as the cold spells, seldom last more than three days, pretty generally in America.

938. *July* 16*th.*—Hot still, but a fine breeze blowing up the river. Not a bit too hot for me, but the natives say it is the hottest weather they recollect in this country; a proof to me that this is a mild climate, as to heat, at any rate. Saw a cat-fish in the market, just caught

out of the river by a hook and line, 4 feet long and eighty pounds weight, offered for 2 dollars. Price of flour, 6 dollars a barrel; fresh beef, 6½ cents, and butter 20 cents per lb.

939. *July* 17*th.*—Set out again, crossing the Ohio into the state of that name, and take the road to Chillicothe, 74 miles from Maysville. Stop about mid-way for the night, travelling over a country generally hilly, and not of good soil, and passing through West Union, a place situated as a town ought to be, upon high and unlevel lands; the inhabitants have fine air to breathe, and plenty of food to eat and drink, and, if they keep their houses and streets and themselves clean, I will ensure them long lives. Some pretty good farms in view of the road, but many abandoned for the richer lands of Indiana and Illinois. Travelling expences much less, hitherto, than in Indiana and some parts of Kentucky; we had plenty of good butter-milk at the farm houses all along the road, free of expence, and the tavern-keepers do not set before us bread made of Indian corn, which we have not yet learned to like very cordially.

940. *July* 18*th.*—Come to Chillicothe, the country improving and more even as we proceed. See some very rich lands on passing Paint Creek, and on approaching the Scioto river; these, like all the *bottom* lands, having

a coat of sediment from their river in addition to the original soil, are by far the richest. Chillicothe is a handsome town, regularly laid out, but, stands upon a flat. I hate the very sight of a level street, unless there be every thing necessary to carry off all filth and water. The air is very fine, so far as it is not contaminated by the pools of water which stand about the town as green as grass. Main sewers, like those at Philadelphia, are much wanted.

941. *July 19th.*—Called upon Mr. Bond, being introduced by letter, and spent a very pleasant evening with him and a large party of his agreeable friends. Left them, much pleased with the society of Chillicothe.

942. *July 20th.*—We were introduced to Governor Worthington, who lives about 2 miles from the town. He took us to his house, and showed us part of his fine estate, which is 800 acres in extent, and all of it elevated table land, commanding an immense view over the flat country in the direction of Lake Erie. The soil is very rich indeed; so rich, that the Governor pointed out a dung heap which was bigger than the barn it surrounded and had grown out of, as a nuisance. The labour of dragging the dung out of the way, would be more than the cost of removing the barn, so that he is actually going to pull the barn down,

and build it up again in another place. This is not a peculiarity of this particular spot of land, for manure has no value here at all. All the stable-dung made at Chillicothe is flung into the river. I dare say, that the Inn we put up at does not tumble into the water less than 300 good loads of horse-dung every year.

943. I had some conversation with Governor Worthington on the subject of domestic manufactures, and was glad to find he is well convinced of the necessity of, or at least of the great benefit that would result from, the general establishment of them in the United States. He has frequently recommended it in his public capacity, he informed me, and I hope he will advocate it with effect. He is a true lover of his country, and no man that I have met with has a more thorough knowledge of the detestable villainy of the odious Boroughmongering government of England, and, of course, it has his full share of hatred.

944. *July 21st.*—Leave Chillicothe. A fine, healthy country and very rich land all the way to New Lancaster, 34 miles from Chillicothe, and 38 from Zanesville. Stop at the house of a German, where we slept, but not in bed, preferring a soft board and something clean for a pillow to a bed of down accompanied with bugs.

945. Nothing remarkable, that I can see, as to the locality of this town of *New Lancaster;* but, the name, alas! it brought to my recollection the horrid deeds done at *Old Lancaster*, the county town of my native county! I thought of *Colonel F———r*, and his conduct towards my poor, unfortunate townsman, Gallant! I thought of the poor, miserable creatures, men, women, and children, who, in the bloody year of 1812, were first instigated by spies to commit arson, and then pursued into death by the dealers in human blood. Amongst the sufferers, upon this particular occasion, there was a boy, who was silly, and who would, at any time, have jumped into a pit for a halfpenny: he was not fourteen years old; and when he was about to be hanged, actually called out for his "*mammy*" to come and save him! Who, that has a heart in his bosom, can help feeling indignation against the cruel monsters! Who can help feeling a desire to see their dreadful power destroyed! The day must come, when the whole of the bloody tragedies of Lancashire will be exposed. In the mean while, here I am in safety from the fangs of the monsters, who oppress and grind my countrymen. The thought of these oppressions, however, I carry about with me; and I cannot help its sometimes bursting forth into words.

946. *July* 22*nd.*—Arrive at Zanesville,* a place finely situated for manufactures, in a nook of the Muskingham, just opposite to the mouth of Licking Creek. It has almost every advantage for manufacturing of all sorts, both as to local situation and as to materials; it excels Wheeling and Steubenville, in many respects, and, in some, even Pittsburgh. The river gives very fine falls near the town, one of them of 12 feet, where it is 600 feet wide; the creek, too, falls in by a fine cascade. What a power for machinery! I should think that as much effect might be produced by the power here afforded as by the united *manual* labour of all the inhabitants of the state. The navigation is very good all the way up to the town, and is now continued round the falls by a canal with locks, so that boats can go nearly close up to Lake Erie. The bowels of the earth afford coal, iron ore, stone, free stone, lime-stone, and *clays:* all of the best, I believe, and the last, the very best yet discovered in this country, and, perhaps, as good as is to be found in any country. All these materials are found in inexhaustible quantities in the hills

---

* For a more particular account of this place, as well, indeed, as of most of the other towns I have visited, see Mr. Mellish's Travels, vol. ii.

and little ridges on the sides of the river and creek, arranged as if placed by the hand of man for his own use. In short, this place has the four elements in the greatest perfection that I have any where yet seen in America. As to manufactures, it is, like Wheeling and Steubenville, nothing in comparison to Pittsburgh.

947. Nature has done her part; nothing is left wanting but machines to enable the people of Ohio to keep their flour at home, instead of exporting it, at their own expence, to support those abroad who are industrious enough to send them back coats, knives, and cups and saucers.

948. *July 23rd.*—All day at Zanesville. Spent part of it very agreeably with Mr. Adams the post-master, and old Mr. Dillon who has a large iron foundery near this.

949. *July 24th.*—Go with Mr. Dillon about 3 miles up the Creek, to see his mills and iron-factory establishment. He has here a very fine water-fall, of 18 feet, giving immense power, by which he works a large iron-forge and foundery, and mills for sawing, grinding, and other purposes.

950. I will here subjoin a list of the prices at Zanesville, of provisions, stock, stores, labour, &c., just as I have it from a resident, whom I can rely upon.

	Dolls.	Cents.	Dolls.	Cents
Flour (superfine), per barrel of 196 lbs. from . . .	5	. to	5	75
Beef, per 100 lbs. . . .	4	. –	4	25
Pork (prime), per 100 lbs.	4	50 –	5	.
Salt, per bushel of 50 lbs.	2	25		
Potatoes, per bushel . .	.	25 –	.	31½
Turnips, ditto	.	20		
Wheat, do. of 60 lbs. to 66 lbs.	.	75		
Indian Corn, ditto, shelled	.	33⅓–	.	50
Oats, ditto	.	25 –	.	33⅓
Rye, ditto	.	50		
Barley ditto	.	75		
Turkeys, of from 12 lbs. to 20 lbs. each	.	37½–	.	50
Fowls	.	12½–	.	18¾
Live Hogs, per 100 lbs. live weight	3	. –	5	.
Cows (the best)	18	. –	25	.
Yoke of Oxen, ditto . .	50	. –	75	.
Sheep	2	50		
Hay, per ton, delivered .	9	. –	10	.
Straw, fetch it and have it.				
Manure, ditto, ditto.				
Coals, per bushel, delivered	.	8		
Butter, per lb. avoirdupois	.	12½–	.	18¾
Cheese, ditto, ditto . . .	.	12½–	.	25
Loaf Sugar	.	50		
Raw ditto	.	31¼		

	Dolls.	Cents.		Dolls.	Cents.
Domestic Raw ditto . .	.	18¾			
Merino Wool, per lb. avoirdupois, washed . . .	1	.			
Three-quarter Merino ditto	.	75			
Common Wool	.	50			
Bricks, per 1000, delivered	6	.	to	7	.
Lime, per bushel, ditto . .	.	18¾			
Sand, in abundance on the banks of the river.					
Glass is sold in boxes, containing 100 square feet; of the common size there are 180 panes in a box, when the price is . .	14	.			
The price rises in proportion to the size of the panes.					
Oak planks, 1 inch thick, per 100 square feet, at the saw-mill.	1	50			
Poplar, the same.					
White Lead, per 100 lbs. delivered	17	.			
Red ditto	17	.			
Litharge	15	.			
Pig Lead	9	50			
Swedish Iron (the best, in bars)	14	.			
Juniatta, ditto, ditto . . .	14	.			

	Dolls.	Cents	Do.lls	Cents.
Mr. Dillon's ditto, ditto .	12	50		
Castings at Mr. Dillon's Foundery, per ton . .	120	.		
Ditto, for machinery, ditto, per lb.	.	8		
Potash, per ton	180	.		
Pearl Ashes, ditto . . .	200	.		
Stone masons and brick-layers, per day, and board and lodging . .	1	50		
Plasterers, by the square yard, they finding themselves in board and lodging and in lime, sand, laths and every thing they use	.	18¾		
Carpenters, by the day, who find themselves and bring their tools	1	25		
Blacksmiths, by the month, and found in board, lodging and tools	30	. to	40	.
Millwrights, per day, finding themselves	1	50–	2	.
Tailors, per week, finding themselves and working 14 or 15 hours a day . .	7	. –	9	.
Shoemakers, the same.				

	Dolls.	Cents	Dolls.	Cents.
Glazier's charge for putting in each pane of glass 8 in. by 10 in. with their own putty and laying on the first coat of paint . . .	.	4 to	.	5
Labourers, per annum, and found	100	. –	120	
The charge of carriage for 100 lbs. weight from Baltimore to Zanesville . .	10	.		
Ditto for ditto by steamboat from New Orleans to Shippingport, and thence by boats, to Zanesville, about	6	50		
Peaches, as fine as can grow, per bushel . . .	.	12½ –	.	25

Apples and Pears proportionably cheaper; sometimes given away, in the country.

951. Prices are much about the same at Steubenville; if any difference, rather lower. If bought in a quantity, some of the articles enumerated might be had a good deal lower. Labour, no doubt, if a job of some length were offered, might be got somewhat cheaper, here.

952. *July 25th.*—Leave Zanesville for Pittsburg, keeping to the United States road; stop at Cambridge, 25 miles. During the first eight miles we met 10 waggons, loaded with emigrants.

953. *July 26th.*—Stop at Mr. Broadshaw's, a very good house on the road, 25 miles from Cambridge. This general government road is by no means well laid out; it goes straight over the tops of the numerous little hills, up and down, up and down. It would have been a great deal nearer in point of time, if not in distance (though I think it would that, too), if a view had been had to the labour of travelling over these everlasting unevennesses.

954. *July 27th.*—To Wheeling in Virginia, 31 miles. They have had tremendous rains in these parts, we hear as we pass along, lately; one of the creeks we came over has overflown so as to carry down a man's house with himself and his whole family. A dreadful catastrophe, but, certainly, one not out of the man's power to have foreseen and prevented; it surprizes me that the people will stick up their houses so near the water's edge. Cross Wheeling Creek several times to-day; it is a rapid stream, and I hope it will not be long before it turns many water-wheels. See much good land, and some pretty good farming.

955. *July 28th.*—Went with a Mr. Graham, a Quaker of this place, who treated us in the most friendly and hospitable manner, to see the new national road from Washington city to this town. It is covered with a very thick layer of nicely broken stones, or stone, rather, laid on with great exactness both as to depth and width, and then rolled down with an iron roller, which reduces all to one solid mass. This is a road made for ever; not like the flint roads in England, rough, nor soft or dirty, like the gravel roads; but, smooth and hard. When a road *is* made in America it is *well* made. An American always plots against labour, and, in this instance, he takes the most effectual course to circumvent it. Mr. Graham took us likewise to see the fine coal mines near this place and the beds of limestone and freestone, none of which I had time to examine as we passed Wheeling in our ark. All these treasures lie very convenient to the river. The coals are principally in one long ridge, about 10 feet wide; much the same as they are at Pittsburgh, in point of quality and situation. They cost 3 cents per bushel to be got out from the mine. This price, as nearly as I can calculate, enables the American collier to earn, upon an average, double the number of cents for the same labour that the collier in England can earn; so that,

as the American collier can, upon an average, buy his flour for one third of the price that the English collier pays for his flour, he receives *six times the quantity of flour for the same labour*. Here is a country for the ingenious paupers of England to come to! They find food and materials, and nothing wanting but their mouths and hands to consume and work them. I should like to see the old toast of the Boroughmongers brought out again; when they were in the height of their impudence their myrmidons used to din in our ears, "Old Eng-"land for ever, and those that do not like her "let them leave her." Let them renew this swaggering toast, and I would very willingly for my part, give another to the same effect for the United States of America. But, no, no! they know better now. They know that they would be taken at their word; and, like the tyrants of Egypt, having got their slaves fast, will (if they can) keep them so. Let them beware, lest something worse than the Red Sea overwhelm them! Like Pharaoh and his Boroughmongers they will not yield to the voice of the people, and, surely, something like, or worse than, their fate shall befall them!

956. They are building a steam-boat at Wheeling, which is to go, they say, 1800 miles up the Missouri river. The wheels are made to work

in the stern of the boat, so as not to come in contact with the floating trees, snaggs, planters,* &c., obstructions most likely very numerous in that river. But, the placing the wheels behind only saves *them;* it is no protection against the *boat's sinking* in case of being pierced by a planter or sawyer.† Observing this, I will suggest a plan which has occurred to me, and which, I think, would provide against sinking, effectually; but, at any rate, it is one which can be tried very easily and with very little expence. —I would make a partition of strong plank; put it in the broadest fore-part of the boat, right across, and put good iron bolts under the bottom of the boat, through these planks, and screw them on the top of the deck. Then put an upright post in the inside of the boat against the middle of the plank partition, and put a spur to the upright post. The partition should be water-tight. I would then load the fore-part of the boat, thus partitioned off, with lumber or such loading as is least liable to injury, and best calculated to stop the progress of a sawyer after it has gone through the boat.—By thus appropriating the fore-part of the boat to the reception of planters and sawyers, it ap-

* Trees tumbled head-long and fixed in the river.

† The same as a planter, only waving up and down.

pears to me that the other part would be secured against all intrusion.

957. *July 29th.*—From Wheeling, through Charlston, changing sides of the river again to Steubenville. My eyes were delighted at Charlston to see the smoke of the coals ascending from the glass-works they have here. This smoke it is that must enrich America; she might save almost all her dollars if she would but bring her invaluable black diamonds into service. Talk of independence, indeed, without coats to wear or knives or plates to eat with!

958. At Steubenville, became acquainted with Messrs. Wills, Ross, and company, who have an excellent and well-conducted woollen manufactory here. They make very good cloths, and at reasonable prices; I am sorry they do not retail them at Philadelphia; I, for one, should be customer to them for all that my family wanted in the woollen-way. Here are likewise a Cotton-mill, a Grist-mill, a Paper-mill, an Iron-foundery and Tan-yards and Breweries. Had the pleasure to see Mr. Wilson, the editor of the Steubenville Gazette, a very public-spirited man, and, I believe, very serviceable to this part of the country. If the policy he so powerfully advocates were adopted, the effects would be grand for America; it would save her dollars while it would help to

draw the nails of the vile Boroughmongers. But, he has to labour against the inveterate effects of the thing the most difficult of all others to move—habit.

959. By what I have been able to observe of this part of the country, those who expect to find what is generally understood by *society*, pretty much the same that they have been accustomed to it on the Atlantic side, or in England, will not be totally disappointed. It is here upon the basis of the same manners and customs as in the oldest settled districts, and it there differs from what it is in England, and here from what it is there, only according to circumstances. Few of the social amusements that are practicable at present, are scarce; dancing, the most rational for every reason, is the most common; and, in an assemblage for this purpose, composed of the farmers' daughters and sons from 20 miles round, an Englishman (particularly if a young one) might very well think his travels to be all a dream, and that he was still in a Boroughmonger country. Almost always the same tunes and dances, same manners, same dress. Ah, it is that same *dress* which is the great evil! It may be a very pretty sight, but, to see the dollars thus danced out of the country into the hands of the Boroughmongers, to the tune of national airs, is a

thing which, if it do not warrant ridicule, will, if America do not, by one unanimous voice, soon put a stop to it.

960. *July 30th.*—From Steubenville, crossing the Ohio for the last time, and travelling through a slip of Virginia and a handsome part of Pennsylvania, to Pittsburgh.

961. *August 1st.*—Sold my horse for 75 dollars, 60 dollars less than I gave for him. A horse changes masters no where so often as in this Western country, and no where so often rises and falls in value. Met a Mr. Gibbs, a native of Scotland, and an old neighbour of mine, having superintended some oil of vitriol works near to my bleach-works on Great Lever, near Bolton, in Lancashire. He now makes oil of vitriol, aquafortis, salts, soap, &c. at this place, and is, I believe, getting rich. Spent a pleasant evening with him.

962. *August 2nd.*—Spent most part of the day with Mr. Gibbs, and dined with him; as the feast was his, I recommended him to observe the latter part of the good Quaker Lady's sermon which we heard at New Albany.

963. *August 3rd.*—Leave Pittsburgh, not without some regret at bidding adieu to so much activity and smoke, for I expect not to see it elsewhere. I like to contemplate the operation by which the greatest effect is produced in a

country. Take the same route and the same stage as on setting out from Philadelphia.

964. *August 4th, 5th, and 6th.*—These three days traversing the romantic Allegany Mountains ; got overturned (a common accident here) *only* once, and then received very little damage: myself none, some of my fellow travellers a few scratches. We scrambled out, and, with the help of some waggoners, set the vehicle on its wheels again, adjusted our "*plunder*" (as some of the Western people call it), and drove on again without being detained more than five minutes. The fourth night slept at Chambersburgh, the beginning of a fine country.

965. *August 7th.*—Travelled over the fine limestone valley before mentioned, and through a very good country all the way, by Little York to Lancaster. Here I met with a person from Philadelphia, who told me a long story about a *Mr. Hulme,* an Englishman, who had brought a large family and considerable property to America. His property, he told me, the said Mr. Hulme had got from the English Government, for the invention of some machine, and that now, having got rich under their patronage, he was going about this country doing the said Government all the mischief he could, and endeavouring to promote the interests of this country. After letting him go on till I was

quite satisfied that he depends mainly for his bread and butter upon the English Treasury, I said, "Well, do you know this Mr. Hulme?" "No, he had only heard of him." "Then I "do, and I know that he never had any patent, "nor ever asked for one, from the English go- "vernment; all he has got he has gained by "his own industry and economy, and, so far "from receiving a fortune from that vile go- "vernment, he had nothing to do with it but "to pay and obey, without being allowed to "give a vote for a Member of Parliament or "for any Government Officer. He is now, "thank God, in a country where he cannot be "taxed but by his own consent, and, if he "should succeed in contributing in any degree "to the downfall of the English Government, "and to the improvement of this country, he "will only succeed in doing his duty." This man could be no other than a dependent of that boroughmongering system which has its feelers probing every quarter and corner of the earth.

966. *August 8th.*—Return to Philadelphia, after a journey of 72 days. My expences for this journey, including every thing, not excepting the loss sustained by the purchase and sale of my horse, amount to 270 dollars and 70 cents.

967. As it is now about a twelvemonth since I have been settled in Philadelphia, or set foot in it, rather, with my family, I will take a look at my books, and add to this Journal what have been the expences of my family for this one year, from the time of landing to this day, inclusive.

	Dolls.		Dolls.	Cents.
House-rent			600	0
Fuel			137	0
Schooling (at day-schools) for my children viz.; for				
Thomas, 14 years of age	40			
Peter and John, ages of 12 and 10, ...	48			
Sarah, 6 years of age, .	18	—	106	0
Boarding of all my family at Mrs. Anthony's Hotel for about a week, on our arrival . . .			30	0
Expences of house-keeping (my family fourteen in number, including two servants) with every other out-going not enumerated above, travelling, incidents, two newspapers a day, &c. &c.			2076	66
Taxes, not a cent.			0	0
Priest, not a cent.			0	0
Total			2999	66

968. "What! nothing to the Parson!" some of my old neighbours will exclaim. No: not a single stiver. The Quakers manage their affairs without Parsons, and I believe they are as good and as happy a people as any religious denomination who are aided and assisted by a Priest. I do not suppose that the Quakers will admit me into their Society; but, in this free country I can form a new society, if I choose, and, if I do, it certainly shall be a Society having a Chairman in place of a Parson, and the assemblage shall discuss the subject of their meeting themselves. Why should there not be as much knowledge and wisdom and common sense, in the heads of a whole congregation, as in the head of a Parson? Ah, but then there are the profits arising from the trade! Some of this holy Order in England receive upwards of 40,000 dollars per annum for preaching probably not more than five or six sermons during the whole year. Well may the Cossack Priests represent Old England as the bulwark of religion! This is the sort of religion they so much dreaded the loss of during the French Revolution; and this is the sort of religion they so zealously expected to establish in America, when they received the glad tidings of the restoration of the Bourbons and the Pope.

END OF THE JOURNAL.

TO

MORRIS BIRKBECK, Esq.

OF

ENGLISH PRAIRIE, ILLINOIS TERRITORY.

North Hempstead, Long Island,
10 *Dec.* 1818.

My Dear Sir,

969. I have read your two little books, namely, the "*Notes on a Journey in America,*" "and the *Letters from the Illinois.*" I opened the books, and I proceeded in the perusal, with *fear and trembling;* not because I supposed it possible for you to put forth an *intended* imposition on the world; but, because I had a sincere respect for the character and talents of the writer; and because I knew how enchanting and delusive are the prospects of enthusiastic minds, when bent on grand territorial acquisitions.

970. My apprehensions were, I am sorry to have it to say, but too well founded. Your books, written, I am sure, without any intention to deceive and decoy, and without any even the smallest tincture of base self-interest, are, in

my opinion, calculated to produce great disappointment, not to say misery and ruin, amongst our own country people (for I will, in spite of your disavowal, still claim the honour of having you for a countryman), and great injury to America by sending back to Europe accounts of that disappointment, misery, and ruin.

971. It is very true, that you decline *advising* any one to go to the ILLINOIS, and it is also true, that your description of the *hardships* you encountered is very candid; but still, there runs throughout the whole of your *Notes* such an account as to the *prospect*, that is to say, the *ultimate effect*, that the book is, without your either wishing or perceiving it, calculated to deceive and decoy. You do indeed describe difficulties and hardships: but, then, you *overcome* them all with so much ease and gaiety, that you make them disregarded by your English readers, who, sitting by their fire-sides, and feeling nothing but the gripe of the Boroughmongers and the tax-gatherer, merely cast a glance at your hardships and fully participate in all your enthusiasm. You do indeed fairly describe the rugged roads, the dirty hovels, the fire in the woods to sleep by, the pathless ways through the wildernesses, the dangerous crossings of the rivers; but, there are the beautiful meadows and rich lands *at last;* there is the

fine freehold domain at the end! There are the giants and the enchanters to encounter; the slashings and the rib-roastings to undergo; but then, there is, *at last*, the lovely languishing damsel to repay the adventurer.

972. The whole of your writings relative to your undertaking, address themselves directly to *English Farmers*, who have property to the amount of two or three thousand pounds, or upwards. Persons of this description are, not by your express words, but by the natural tendency of your writings, *invited*, nay, strongly invited, to emigrate with their property to the Illinois Territory. Many have already acted upon the invitation. Many others are about to follow them. I am convinced, that their doing this is unwise, and greatly injurious, not only to them, but to the character of America as a country to emigrate to, and, as I have, in the first Part of this work, promised to give, as far as I am able, a *true* account of America, it is my duty to state the *reasons* on which this conviction is founded; and, I address the statement to you, in order, that, if you find it erroneous, you may, in the like public manner, show wherein I have committed error.

973. We are speaking, my dear Sir, of English Farmers possessing each two or three thousand pounds sterling. And, before we

proceed to inquire, whether such persons ought to emigrate to the *West* or to the *East*, it may not be amiss to inquire a little, whether they ought to *emigrate at all!* Do not start, now! For, while I am very certain that the emigration of *such persons* is not, in the end, calculated to produce benefit to America, as a nation, I greatly doubt of its being, *generally speaking*, of any benefit to the emigrants themselves, if we take into view the chances of their speedy relief at home.

974. Persons of advanced age, of settled habits, of deep rooted prejudices, of settled acquaintances, of contracted sphere of movement, do not, to use Mr. George Flower's expression, "*transplant well.*" Of all such persons, Farmers transplant worst; and, of all Farmers, English Farmers are the worst to transplant. Of some of the *tears*, shed in the Illinois, an account reached me several months ago, through an eye-witness of perfect veracity, and a very sincere friend of freedom, and of you, and whose information was given me, unasked for, and in the presence of several Englishmen, every one of whom, as well as myself, most ardently wished you success.

975. It is nothing, my dear Sir, to say, as you do, in the Preface to the *Letters from the Illinois*, that, "as little would I encourage the

" emigration of the tribe of *grumblers*, people
" who are petulant and discontented under the
" *every-day* evils of life. Life has its petty
" miseries in *all situations* and climates, to be
" mitigated or cured by the continual efforts of
" an elastic spirit, or to be borne, if incurable,
" with cheerful patience. But the *peevish emi-*
" *grant* is perpetually comparing the *comforts*
" he has quitted, but never could enjoy, with
" the *privations of his new allotment*. He over-
" looks the *present good*, and broods over the
" evil with *habitual perverseness;* whilst in the
" recollection of the past, he dwells on the
" good only. Such people are always *bad as-*
" *sociates*, but they are *an especial nuisance* in
" an infant colony."

976. Give me leave to say, my dear Sir, that there is too much *asperity* in this language, considering who were the objects of the censure. Nor do you appear to me to afford, in this instance, a very happy illustration of the absence of that *peevishness*, which you perceive in others, and for the yielding to which you call them a *nuisance;* an appellation much too harsh for the object and for the occasion. If you, with all your elasticity of spirit, all your ardour of pursuit, all your compensations of fortune in prospect, and all your gratifications of fame in possession, cannot with patience hear the wail-

ings of some of your neighbours, into what source are they to dip for the waters of content and good-humour?

977. It is no "*every-day evil*" that they have to bear. For an English Farmer, and, more especially, an English Farmer's wife, after crossing the sea and travelling to the Illinois, with the consciousness of having expended a third of their substance, to purchase, as yet, nothing but sufferings; for such persons to boil their pot in the gipsy-fashion, to have a mere board to eat on, to drink whisky or pure water, to sit and sleep under a shed far inferior to their English cow-pens, to have a mill at twenty miles distance, an apothecary's shop at a hundred, and a doctor no where: these, my dear Sir, are not, to *such people*, "*every-day* evils of "life." You, though in your little "cabin," have your *books*, you have your name circulating in the world, you have it to be given, by and bye, to a city or a county; and, if you fail of brilliant success, you have still a sufficiency of fortune to secure you a safe retreat. Almost the whole of your neighbours must be destitute of all these sources of comfort, hope, and consolation. As they *now are*, their change is, and must be, for the worse; and, as *to the future*, besides the uncertainty attendant, every where, on that which is to come, they ought to be ex-

cused, if they, at their age, despair of seeing days as happy as those that they have seen.

978. It were much better for *such people* not to emigrate at all; for while they are *sure* to come into a state of some degree of suffering, they leave behind them the *chance* of happy days; and, in my opinion, a *certainty* of such days. I think it next to impossible for any man of tolerable information to believe, that the present tyranny of the seat-owners can last another two years. As to *what change* will take place, it would, perhaps, be hard to say: but, that *some great change* will come is certain; and, it is also certain, that the change *must be for* the better. Indeed, one of the motives for the emigration of many is said to be, that they think a *convulsion* inevitable. Why should such persons as I am speaking of fear a convulsion? Why should they suppose, that they will suffer by a convulsion? What have *they* done to provoke the rage of the blanketteers? Do they think that their countrymen, all but themselves, will be transformed into prowling wolves? This is precisely what the Boroughmongers wish them to believe; and, believing it, they *flee* instead of remaining to assist to keep the people down, as the Boroughmongers wish them to do.

979. Being here, however, they, as you say, *think only of the good* they have left behind

them, and of *the bad they find here.* This is no fault of theirs: it is the natural course of the human mind; and this you ought to have known. You yourself acknowledge, that England "*was never so dear to you as it is now in* "*recollection:* being no longer under its base "oligarchy, I can think of my native country "and her *noble institutions*, apart from her *poli-* "*tics.*" I may ask you, by the way, what *noble institutions* she has, which are not of a *political nature?* Say the *oppressions of her tyrants*, say that you can think of her and love her renown and her famous political institutions, apart from those oppressions, and then I go with you with all my heart; but, so thinking, and so feeling, I cannot say with you, in your NOTES, that England is to me "*matter of history*," nor with you, in your LETTERS FROM THE ILLINOIS, that "where *liberty* is, there is *my country*."

980. But, leaving this matter, for the present, if English Farmers must emigrate, why should they encounter *unnecessary* difficulties? Coming from a country like a garden, why should they not stop in another *somewhat resembling* that which they have lived in before? Why should they, at an expence amounting to a large part of what they possess, prowl two thousand miles at the hazard of their limbs and lives, take women and children through scenes of hardship

and distress not easily described, and that too, to live like gipsies at the end of their journey, for, at least, a year or two, and, as I think I shall show, without the smallest chance of their *finally* doing so well as they may do in these Atlantic States? Why should an English Farmer and his family, who have always been jogging about a snug home-stead, eating regular meals, and sleeping in warm rooms, push back to the Illinois, and encounter those hardships, which require all the habitual disregard of comfort of an American back-woods-man to overcome? Why should they do this? The undertaking is hardly reconcileable to reason in an Atlantic *American* Farmer who has half a dozen sons, all brought up to use the axe, the saw, the chisel and the hammer from their infancy, and every one of whom is ploughman, carpenter, wheelwright and butcher, and can work from sun-rise to sun-set, and sleep, if need be, upon the bare boards. What, then, must it be in an English Farmer and his family of helpless mortals? Helpless, I mean, in this scene of such novelty and such difficulty? And what is his *wife* to do; she who has been torn from all her relations and neighbours, and from every thing that she liked in the world, and who, perhaps, has never, in all her life before, been *ten miles* from the cradle in which she was nursed? An

American farmer mends his plough, his waggon, his tackle of all sorts, his household goods, his shoes; and, if need be, he *makes* them all. Can our people do all this, or any part of it? Can they live without bread for months? Can they live without beer? Can they be otherwise than miserable, cut off, as they must be, from all intercourse with, and hope of hearing of, their relations and friends? The truth is, that this is not *transplanting*, it is *tearing up and flinging away*.

981. *Society!* What society can these people have? 'Tis true they have nobody to envy, for nobody can have any thing to enjoy. But there may be, and there must be, mutual complainings and upbraidings; and every unhappiness will be traced directly to him who has been, however unintentionally, the cause of the unhappy person's removal. The very foundation of your plan necessarily contained the seeds of discontent and ill-will. A *colony* all from the same country was the very worst project that could have been fallen upon. You took upon yourself the *charge* of MOSES without being invested with any part of his *authority;* and absolute as this was, he found the charge so heavy, that he called upon the Lord to share it with him, or to relieve him from it altogether. Soon after you went out, an Unitarian Priest, upon my asking

what you were going to do in that wild country, said, you were going to form a community, who would be " content to worship *one God.*" " I hope not," said I, " for he will have plagues " enough without adding a priest to the number." But, perhaps, I was wrong: for AARON was of great assistance to the leader of the Israelites.

982. As if the inevitable effects of disappointment and hardship were not sufficient, you had, too, a sort of *partnership* in the *leaders.* This is *sure* to produce feuds and bitterness in the long run. Partnership-sovereignties have furnished the world with numerous instances of poisonings and banishments and rottings in prison. It is as much as merchants, who post their books every Sunday, can do to get along without quarrelling. Of man and wife, though they are flesh of flesh and bone of bone, the harmony is not always quite perfect, except in France, where the husband is the servant, and in Germany and Prussia, where the wife is the slave. But, as for a partnership sovereignty without disagreement, there is but one single instance upon record; that, I mean, was of the *two kings of Brentford,* whose cordiality was, you know, so perfect, that they both smelt to the same nosegay. This is, my dear Sir, no bantering. I am quite serious. It is impossible

that separations should not take place, and equally impossible that the neighbourhood should not be miserable. This is not the way to settle in America. The way is, to go and *sit yourself down amongst the natives.* They are already settled. They can *lend* you what you want to borrow, and happy they are always to do it. And, which is the great thing of all great things, you have their *women* for *your women to commune with!*

983. RAPP, indeed, has done great things; but RAPP has the authority of Moses and that of Aaron united in his own person. Besides, Rapp's community observe in reality that celibacy, which Monks and Nuns pretend to, though I am not going to take my oath, mind, that none of the tricks of the Convent are ever played in the tabernacles of *Harmony.* At any rate, Rapp secures the *effects* of celibacy; first, an absence of the expence attending the breeding and rearing of children, and, second, unremitted labour of woman as well as man. But, where, in all the world is the match of this to be found? Where else shall we look for a Society composed of persons willing and able to forego the gratification of the most powerful propensity of nature, for the sake of getting money together? Where else shall we look for a band of men and women who love money

better than their own bodies? Better than their *souls* we find people enough to love money; but, who ever before heard of a set that preferred the love of money to that of their bodies? Who, before, ever conceived the idea of putting a stop to the procreation of children, for the sake of saving the expence of bearing and breeding them? This Society, which is a perfect prodigy and monster, ought to have the image of MAMMON in their place of worship; for that is the object of their devotion, and not the God of nature. Yet the persons belonging to this unnatural association are your nearest neighbours. The masculine things here, called women, who have imposed barrenness on themselves, out of a pure love of gain, are the nearest neighbours of the affectionate, tender-hearted wives and mothers and daughters, who are to inhabit your colony, and who are, let us thank God, the very reverse of the petticoated Germans of Harmony.

984. In such a situation, with so many circumstances to annoy, what happiness can an English family enjoy in that country, so far distant from all that resembles what they have left behind them? " The fair Enchantress, " *Liberty*," of whom you speak with not too much rapture, they would have found in any of *these States*, and, in a garb, too, by which

they would have *recognised* her. Where they now are, they are *free* indeed; but their freedom is that of the wild animals in your woods. It is not *freedom*, it is *no government*. The GIPSIES, in England, are *free;* and any one, who has a mind to live in a cave, or cabin, in some hidden recess of our *Hampshire forests*, may be *free* too. The English farmer, in the Illinois, is, indeed, beyond the reach of the Boroughmongers; and so is the man that is in the grave. When it was first proposed, in the English Ministry, to *drop quietly* the title of *King of France* in the enumeration of our king's titles, and, when it was stated to be an expedient *likely to tend to a peace*, Mr. WINDHAM, who was then a member of the Cabinet, said: " As this is a measure of *safety*, and as, " doubtless, we shall hear of others of the same " cast, what think you *of going under ground* " *at once?*" It was a remark enough to cut the liver out of the hearers; but Pitt and his associates had no livers. I do not believe, that any twelve Journeymen, or Labourers, in England would have voted for the adoption of this mean and despicable measure.

985. If, indeed, the Illinois were the *only* place out of the reach of the Borough-grasp; and, if men are resolved to get out of that reach; then, I should say, Go to the Illinois,

by all means. But, as there is a country, a settled country, a free country, full of kind neighbours, full of all that is good, and when this country is to be *traversed* in order to *get at* the acknowledged hardships of the Illinois, how can a sane mind lead an English Farmer into the expedition?

986. It is the enchanting damsel that makes the knight encounter the hair-breadth scapes, the sleeping on the ground, the cooking with cross-sticks to hang the pot on. It is the *Prairie*, that pretty French word, which means green grass bespangled with daisies and cowslips! Oh, God! What delusion! And that a man of sense; a man of superior understanding and talent; a man of honesty, honour, humanity, and lofty sentiment, should be the cause of this delusion; I, my dear Sir, have seen *Prairies* many years ago, in America, as fine as yours, as fertile as yours, though not so extensive. I saw those Prairies settled on by American Loyalists, who were carried, with all their goods and tools to the spot, and who were furnished with four years' provisions, all *at the expence of England;* who had the lands *given them;* tools *given them;* and who were thus seated down on the borders of *creeks,* which gave them easy communication with the inhabited plains near the sea. The

settlers that I particularly knew were Connecticut men. Men with families of sons. Men able to do as much in a day at the works necessary in their situation as so many Englishmen would be able to do in a week. They began with a *shed;* then rose to a *log-house;* and next to a *frame-house;* all of their own building. I have seen them manure their land with *Salmon* caught in their creeks, and with *pigeons* caught on the land itself. It will be a long while before you will see such beautiful *Corn-fields* as I saw there. Yet nothing but the danger and disgrace which attended their return to Connecticut *prevented their returning*, though there they must have begun the world anew. I saw them in their log-huts, and saw them in their frame-houses. They had overcome all their difficulties as settlers; they were under a government which required neither tax nor service from them; they were as happy as people could be as to ease and plenty; but, still, they *sighed for Connecticut;* and especially the *women*, young as well as old, though we, gay fellows with worsted or silver lace upon our bright red coats, did our best to make them happy by telling them entertaining stories about Old England, while we drank their coffee and grog by gallons, and eat their fowls, pigs and sausages and sweet-meats, by wheel-barrow loads;

for, though we were by no means *shy*, their hospitality far exceeded our appetites. I am an old hand at the work of settling in wilds. I have, more than once or twice, had to begin my nest and go in, like a bird, making it habitable by degrees; and, if I, or, if such people as my old friends above-mentioned, with every thing found for them and brought to the spot, had difficulties to undergo, and *sighed for home* even after all the difficulties were over, what must be the lot of an English Farmer's family in the Illinois?

987. All this I told you, my dear sir, in London just before your departure. I begged of you and Mr. Richard Flower both, not to think of the Wildernesses. I begged of you to go to within a day's ride of some of these great cities, where your ample capital and your great skill could not fail to place you upon a footing, at least, with the richest amongst the most happy and enlightened Yeomanry in the world; where you would find every one to praise the improvements you would introduce, and nobody to envy you any thing that you might acquire. Where you would find society as good, in all respects, as that which you had left behind you. Where you would find neighbours ready prepared for you far more generous and hospitable than those in England *can* be,

loaded and pressed down as they are by the inexorable hand of the Borough-villains. I offered you a letter (which, I believe, I sent you), to my friends the PAULS. " But," said I, " you want no letter. Go into Philadelphia, " or Bucks, or Chester, or Montgomery Coun- " ty; tell any of the Quakers, or any body " else, that you are an English Farmer, come " to settle amongst them; and, I'll engage that " you will instantly have friends and neigh- " bours as good and as cordial as those that " you leave in England."

988. At this very moment, if this plan had been pursued, you would have had a beautiful farm of two or three hundred acres. Fine stock upon it feeding on Swedish Turnips. A house overflowing with abundance; comfort, ease, and, if you chose, elegance, would have been your inmates; *libraries*, public and private within your reach; and a communication with England much more quick and regular than that which you now have even with Pittsburgh.

989. You say, that " Philadelphians *know* " *nothing* of the Western Countries." Suffer me, then, to say, that you know nothing of the *Atlantic States*, which, indeed, is the only apology for your saying, that the *Americans have no mutton fit to eat*, and regard it *only as a thing fit for dogs*. In this island *every* farmer

has sheep. I kill *fatter* lamb than I ever *saw* in England, and the *fattest* mutton I ever saw, was in company with Mr. Harline, in Philadelphia market last winter. At BRIGHTON, near Boston, they produced, at a cattle shew this fall, an ox of *two-thousand seven-hundred pounds* weight, and sheep much finer, than you and I saw at the Smithfield Show in 1814. Mr. Judge Lawrence of this county, has kept, for seven years, an average of *five hundred Merinos* on his farm of *one hundred and fifty acres*, besides raising twenty acres of Corn and his usual pretty large proportion of grain! Can your Western Farmers beat that? Yes, in extent, as the surface of five dollars beats that of a guinea.

990. I suppose that Mr. Judge Lawrence's farm, close by the side of a bay that gives him two hours of water carriage to New-York; a farm with twenty acres of meadow, *real prairie;* a gentleman's house and garden; barns, sheds, cider-house, stables, coach-house, corn-cribs, and orchards that may produce from four to eight thousand bushels of apples and pears: I suppose, that this farm is worth *three hundred dollars an acre:* that is, forty-five thousand dollars; or about, *twelve or thirteen thousand pounds.*

991. Now, then, let us take a look at your

estimate of the expences of *sitting down* in the prairies.

Copy from my Memorandum Book.

992. Estimate of money required for the comfortable establishment of my family on Bolting House, now English, prairie; on which the first instalment is paid. About 720 acres of woodland, and 720 prairie—the latter to be chiefly grass:—

		Dollars.
Second instalment, August, 1819,	720	
Third ditto . . . August, 1820,	720	
Fourth ditto . . August, 1821,	720	
	——	2,160
Dwelling-house and appurtenances . .		4,500
Other buildings		1,500
4680 rods of fencing, viz. 3400 on the prairie, and 1280 round the woodland		1,170
Sundry wells, 200 dollars; gates, 100 dollars; cabins, 200 dollars . . .		500
100 head of cattle, 900 dollars; 20 sows, &c. 100 dollars; sheep, 1000 dollars .		2,000
Ploughs, waggons, &c. and sundry tools and implements		270
Housekeeping until the land supplies us		1,000
Carried over . . .		13,100

	Dollars.
Brought over	13,100
Shepherd one year's wages, herdsmen one year, and sundry other labourers	1,000
One cabinet-maker, one wheel-wright, one year, making furniture and implements, 300 dollars each	600
Sundry articles of furniture, ironmongery, pottery, glass, &c.	500
Sundries, fruit trees, &c.	100
First instalment already paid	720
Five horses on hand, worth	300
Expence of freight and carriage of linen, bedding, books, clothing, &c . . .	1,000
Value of articles brought from England	4,500
Voyage and journey	2,000
	Doll. 23,820

23,820 dollars = £5,359 sterling.

Allow about 600 dollars more for seed and corn	141
	£5,500

993. So, here is more than one third of the amount of Mr. Judge Lawrence's farm. To be sure, there are only about 18,000 dollars expended on land, buildings, and *getting at them;* but, *what a life* is that which you are to lead for *a thousand dollars a year*, when two good domestic servants will cost *four hundred*

of the money? Will you live like one of the Yeomen of your rank *here?* Then, I assure you, that your domestics and groceries (the latter three times as dear as they are here) and crockery-ware (equally dear) will more than swallow up that pitiful sum. You allow six thousand dollars for *buildings.* Twice the sum would not put you, in this respect, upon a footing with Mr. Lawrence. His land is all completely fenced and his grain in the ground. His apple trees have *six thousand bushels of apples in their buds,* ready to come out in the spring; and, a large part of these to be sold at a high price to go on ship-board. But, what is to give you his *market?* What is to make your pork, as soon as killed, sell for 9 or 10 dollars a hundred, and your cows at 45 or 50 dollars each, and your beef at 7 or 8 dollars a hundred, and your corn at a dollar, and wheat at two dollars a bushel?

994. However, happiness is in the *mind;* and, if it be necessary to the gratification of your mind to inhabit a wilderness and be the owner of a large tract of land, you are right to seek and enjoy this gratification. But, for the plain, plodding *English Farmer,* who simply seeks safety for his little property, with some addition to it for his children; for such a person to *cross* the Atlantic states in search of

safety, tranquillity and gain in the Illinois, is, to my mind, little short of madness. Yet, to this mad enterprize is he allured by your captivating statements, and which statements become decisive in their effects upon his mind, when they are reduced to *figures*. This, my dear Sir, is the part of your writings, which has given me most pain. You have not *meant to deceive*; but you have first practised a deceit upon yourself, and then upon others. All the disadvantages you *state*; but, then, you accompany the statement by telling us how *quickly* and how *easily* they will be *overcome*. Salt, Mr. HULME finds, even at ZANESVILLE, at *two dollars and a half a bushel*; but, you tell us, that it *soon will be* at three quarters of a dollar. And thus it goes all through.

995. I am happy, however, that you have given us *figures* in your account of what an English farmer may do *with two thousand pounds*. It is alluring, it is fallacious, it tends to disappointment, misery, ruin and broken hearts; but it is open and honest in intention, and it affords us the means of detecting and exposing the fallacy. Many and many a family have returned to New England after having emigrated to the West in search of *fine estates*. They, able workmen, exemplary livers, have returned to labour in their native States amongst

their relations and old neighbours; but, what are our poor ruined countrymen to do, when they become pennyless? If I could root my country from my heart, common humanity would urge me to make an humble attempt to dissipate the charming delusions, which have, without your perceiving it, gone forth from your sprightly and able pen, and which delusions are the more dangerous on account of your justly high and well-known character for understanding and integrity.

996. The statement, to which I allude, stands as follows, in your *tenth Letter* from *the Illinois.*

997. A capital of 2000*l.* sterling, (8,889 dollars,) may be invested on a section of such land, in the following manner, *viz.*

	Dollars.
Purchase of the land, 640 acres, at 2 dollars per acre	1280
House and buildings, exceedingly convenient and comfortable, may be built for	1500
A rail fence round the woods, 1000 rods, at 25 cents per rod	250
About 1800 rods of ditch and bank, to divide the arable land into 10 fields .	600
Planting 1800 rods of live fence . . .	150
Carried over . .	3780

	Dollars.
Brought over . . .	3780
Fruit trees for orchard, &c.	100
Horses and other live stock	1500
Implements and furniture	1000
Provision for one year, and sundry incidental charges	1000
Sundry articles of linen, books, apparel, implements, &c. brought from England	1000
Carriage of ditto, suppose 2000 lbs. at 10 dollars per cwt.	200
Voyage and travelling expences of one person, suppose	309
	8889

Note.—The first instalment on the land is 320 dollars, therefore 960 dollars of the purchase money remain in hand to be applied to the expences of cultivation, in addition to the sums above stated.

Expenditure of first Year.

Breaking up 100 acres, 2 dollars per acre	200
Indian corn for seed, 5 barrels, (a barrel is five bushels)	10
Planting ditto	25
Carried over . . .	235

	Dollars.
Brought over . . .	235
Horse-hoeing ditto, one dollar per acre .	100
Harvesting ditto, 1½ dollar per acre . .	150
Ploughing the same land for wheat, 1 dollar per acre	100
Seed wheat, sowing and harrowing . .	175
Incidental expences	240
	1000

Produce of first Year.	
100 acres of Indian corn, 50 bushels (or 10 barrels) per acre, at 2 dollars per barrel	2000
Net produce	1000

Expenditure of second Year.	
Breaking up 100 acres for Indian corn, with expences on that crop	485
Harvesting and threshing wheat, 100 acres	350
Ploughing 100 acres for wheat, seed, &c.	275
Incidents	290
	1400

Produce of second Year.		
100 acres Indian corn, 10 barrels per acre, 2 dollars per barrel	2000	
100 acres wheat, 20 bushels per acre, 75 dollars per barrel . .	1500	—3500
Net produce		2100

Expenditure of third Year.		Dollars.
Breaking up 100 acres as before, with expences on crop of Indian corn . .		485
Ploughing 100 acres of wheat stubble for Indian corn		100
Horse hoeing, harvesting, &c. ditto . .		285
Harvesting and threshing 100 acres wheat		350
Dung-carting 100 acres for wheat, after second crop of Indian corn		200
Ploughing 200 acres wheat, seed, &c. .		550
Incidents		330
		2300
Produce of third Year.		
200 acres of Indian corn, 10 barrels per acre, 2 dollars per barrel	4000	
100 acres wheat, 20 bushels per acre, 75 dollars per barrel . .	1500—	5500
	Net produce	3200
Expenditure of fourth Year.		
As the third		2300
Harvesting and threshing 100 acres more wheat		350
Additional incidents		50
		2700

Produce of fourth Year.		Dollars.
200 acres Indian corn, as above .	4000	
200 acres wheat	3000—	7000
	Net produce	4300

Summary.

	EXPENCES. Dollars.		PRODUCE. Dollars.	
First year . . .	1000	. .	2000	
Second	1400	. .	3500	
Third	2300	. .	5500	
Fourth	2700	. .	7000	
			18000	
House-keeping and other expences for four years . . .	4000		11,400	
Net proceeds per annum				1650
Increasing value of land by cultivation and settlements, half a dollar per ann. on 640 acres				320
Annual clear profit				1970

998. "Twenty more: kill 'em! Twenty "more: kill them too!" No: I will not compare you to BOBADIL: for he was an intentional deceiver; and you are unintentionally deceiving others and yourself too. But, really, there is in this statement something so extravagant; so perfectly wild; so ridiculously and staringly

untrue, that it is not without a great deal of difficulty that all my respect for you personally can subdue in me the temptation to treat it with the contempt due to its intrinsic demerits.

999. I shall notice only a few of the items. A house, you say, "*exceedingly* convenient and "comfortable, together with farm-buildings, "may be built for 1500 dollars." Your own *intended* house you estimate at 4500, and your out-buildings at 1500. So that, *if* this house of the farmer (an English farmer, mind) and his buildings, are to be "*exceedingly convenient* "and *comfortable*," for 1500 dollars, your house and buildings must be on a scale, which, if not perfectly *princely*, must savour a good deal of aristocratical distinction. But, this *if* relieves us; for even your house, built of pine timber and boards, and covered with cedar shingles, and finished only as a *good plain farm-house* ought to be, will, if it be *thirty-six feet front, thirty-four feet deep*, two rooms in front, kitchen, and wash-house behind, four rooms above, and a cellar beneath; yes, this house alone, the bare empty house, with doors and windows suitable, will cost you more than *six thousand dollars.* I state this upon good authority. I have taken the estimate of a building carpenter. "What "Carpenter?" you will say. Why, a Long Island carpenter, and the house to be built

within a mile of Brooklyn, or two miles of New York. And this is giving you all the advantage, for here the pine is cheaper than with you; the shingles cheaper; the lime and stone and brick as cheap or cheaper; the glass, iron, lead, brass and tin, all at half or a quarter of the Prairie price: and, as to *labour*, if it be not cheaper here than with you, men would do well *not to go so far in search of high wages!*

1000. Let no simple Englishman imagine, that here, at and near New York, in this *dear place*, we have to pay for the boards and timber *brought from a distance;* and that you, the happy people of the land of daisies and cowslips, can cut down *your own good and noble oak trees upon the spot, on your own estates, and turn them into houses without any carting*. Let no simple Englishman believe such idle stories as this. To dissipate all such notions, I have only to tell him, that the American farmers on this island, when they have buildings to make or repair, go and *purchase* the pine timber and boards, at the very same time that they *cut down their own oak trees and cleave up and burn them as fire-wood!* This is the universal practice in all the parts of America that I have ever seen. What is the cause? Pine wood is *cheaper*, though *bought*, than the oak is *without buying*. This fact, which nobody can deny, is a complete proof that you gain no advantage from being in woods, as far as *building* is concerned. And the truth is, that the boards and plank, which have been used in the Prairie, *have actually been brought from the Wabash*, charged with ten miles *rough land carriage:* how far they may have come down the Wabash I cannot tell.

1001. Thus, then, the question is settled that building must be *cheaper here than in the Illinois.* If, therefore, a house, 36 by 34 feet, cost *here* 6000 dollars, what can a man get *there* for 1500 dollars? A miserable hole, and no more. But, here are to be *farm-buildings and all*, in the 1500 dollars' worth! A barn, 40 feet by 30, with floor, and with stables in the sides, cannot be built for 1500 dollars, leaving out waggon-house, corn-crib, cattle-hovels, yard fences, pig-sties, smoke house, and a great deal more! And yet, you say, that all these, and a farm-house into the bargain, all "*exceedingly comfortable* "and *convenient*," may be had for 1500 dollars!

1002. Now, you know, my dear Sir, that this is said in the face of all America. Farmers are my readers. They *all* understand these matters. They are not only good, but impartial judges; and I call upon you to contradict, or even question, my statements, if you can.

1003. Do my eyes deceive me? Or do I really see *one hundred and fifty dollars* put down as the expence of "*planting one thousand eight "hundred rod of live fence*"? That is to say, *nine cents, or four pence half-penny sterling, a rod! What* plants? *Whence* to come? Drawn out of the woods, or first sown in a nursery? Is it *seed* to be sown? *Where* are the seeds to come from? No levelling of the top of the bank; no drill; no sowing; no keeping clean for a year or two: or, *all these for nine cents a rod*, when the same works cost *half a dollar a rod in England!*

1004. *Manure* too! And do you really want manure then? And, where, I pray you, are you to get manure for 100 acres? But, supposing you to have it, do you seriously mean to tell us that you will carry it on for two dollars an acre? The *carrying on*, indeed, might perhaps be done for that, but, who pays for the *filling* and for the *spreading?* Ah! my dear Sir, I can well imagine your feelings at putting down the item of dung-carting, trifling as you make it appear upon paper. You now recollect my words when I last had the pleasure of seeing you, in Catherine Street, a few days before the departure of us both. I then dreaded the dung-cart, and recommended the Tullian System to you, by which you would have the same crops every year, without manure; but, unfortunately for my advice, you sincerely believed your land would be already too rich, and that your main difficulty would be, not to *cart on* manure, but to *cart off* the produce!

1005. After this, it appears unnecessary for me to notice any other part of this Transalleganian romance, which I might leave to the admiration of the Edinburgh Reviewers, whose knowledge of these matters is quite equal to what they have discovered as to the Funding System and Paper Money. But when I think of the flocks of poor English Farmers, who are tramping away towards an imaginary, across a real land of milk and honey, I cannot lay down the pen, till I have noticed an item or two of the *produce.*

1006. The farmer is to have 100 acres of Indian corn, the first year. The minds of you gentlemen who cross the Allegany seem to expand, as it were, to correspond with the extent of the horizon that opens to your view; but, I can assure you, that if you were to talk to a farmer on this side of the mountains of a field of Corn of a hundred acres during the first year of a settlement, with grassy land and hands scarce, you would frighten him into a third-day ague. In goes your Corn, however! "Twenty "more: kill 'em!" Nothing but ploughing: no harrowing; no marking; and only a horse-hoeing, during the summer, at *a dollar an acre.*

The planting is to cost only *a quarter of a dollar an acre.* The planting will cost a *dollar an acre.* The horse-hoeing in your grassy land, *two dollars.* The *hand-hoeing*, which must be *well* done, or you will have no corn, *two dollars;* for, in spite of your teeth, your rampant natural grass will be up before your corn, and a man must go to *a thousand hills* to do *half an acre a day.* It will cost *two dollars* to harvest a hundred bushels of *corn ears.* So that here are about 400 dollars of expences on the Corn alone, to be added. A *trifle*, to be sure, when we are looking through the Transalleganian glass, which diminishes out-goings and magnifies incomings. However, here are four hundred dollars.

1007. In goes the plough for wheat? "In "him again! Twenty more!" But, this is in *October*, mind. Is the Corn off? It may be; but, where are the *four hundred waggon loads of corn stalks?* A prodigiously fine thing is this forest of fodder, as *high* and as *thick* as an English coppice. But, though it be of *no use to you*, who have the *meadows* without bounds, this coppice must be *removed*, if you please, before you plough for wheat!

1008. Let us pause here, then; let us look at the *battalion*, who are at work; for, there must be little short of a Hessian Battalion. Twenty men and twenty horses *may* husk the Corn, cut and cart the stalks, plough and sow and harrow for the wheat; twenty two-legged and twenty four-legged animals *may* do the work in the proper time; but, if they do it, they must work *well.* Here is a goodly group to look at, for an English Farmer, without a penny in his pocket; for all his money is *gone long ago*, even according to your own estimate; and, here, besides the expence of cattle and tackle, are 600 dollars, in bare wages, to be paid in a month! You and I both have forgotten the *shelling* of the Corn, which, and putting it up, will come to 50 dollars more at the least, leaving the price of the barrel to be paid for by the purchaser of the Corn.

1009. But, what did I say? *Shell* the Corn? It must go into the *Cribs* first. It cannot be shelled *immediately*. And it must not be thrown into *heaps*. It must be put into *Cribs*. I have had made out an estimate of the expence of the Cribs for *ten thousand bushels* of Corn Ears: that is the crop; and the Cribs will cost 570 dollars! Though, mind, the farmer's *house, barns, stables, waggon-house*, and all, are to cost but 1500 dollars! But, the third year, our poor simpleton is to have 200 acres of corn! "Twenty "more: kill 'em!" Another 570 dollars for Cribs!

1010. However, crops now come tumbling on him so fast, that he must struggle hard not to be stifled with his own superabundance. He has now got 200 acres of corn and 100 acres of wheat, which latter he has, indeed, had one *year before!* Oh, madness! But, to proceed. To get in these crops and to sow the wheat, first taking away 200 acres of *English coppice* in stalks, will, with the *dunging* for the wheat, require, at least, *fifty good men*, and *forty good horses or oxen*, for *thirty days*. Faith! when farmer Simpleton sees all this (in his *dreams* I mean), he will think himself a farmer of the rank of JOB, before Satan beset that example of patience, so worthy of imitation, and so seldom imitated.

1011. Well, but Simpleton must bustle to *get in* his wheat. *In*, indeed! What can *cover* it, but the canopy of heaven? A *barn!* It will, at *two English waggon loads of sheaves to an acre*, require a barn a hundred feet long, fifty feet wide, and twenty-three feet high up to the eaves; and this barn, with two proper floors, will cost more than *seven thousand dollars*. He will put it in *stacks;* let him add six men to his battalion then. He *will thrash it in the field;* let him add ten more men! Let him, at once, send and press the Harmonites into his service, and make RAPP march at their head, for, never will he by any other means get in the crop; and, even then, if he pay fair wages, he will lose by it.

1012. After the crop is in and the seed sown, in the fall, what is to become of Simpleton's men till Corn ploughing and planting time in the spring? And, then, when the planting is done, what is to become of them till harvest time? Is he, like BAYES, in the Rehearsal, to lay them down when he pleases, and when he pleases make them rise up again? To hear you talk about these crops, and, at other times to hear you advising others to bring labourers from England, one would think you, for your own part, able, like CADMUS, to make men start up out of the earth. How would one ever have thought it possible for infatuation like this to seize hold of a mind like yours?

1013. When I read in your Illinois Letters, that you had *prepared* horses, ploughs, and other things, *for putting in a hundred acres of Corn in the Spring*, how I pitied you! I saw all your plagues, if you could not see them. I saw the grass choking your plants; the grubs eating them; and you fretting and turning from the sight with all the pangs of sanguine baffled hope. I expected you to have *ten bushels*, instead of *fifty*, upon an acre. I saw your confusion, and participated in your mortification.

From these feelings I was happily relieved by the Journal of our friend HULME, who informs the world, and our countrymen in particular, that you had not, in *July last, any Corn at all growing!*

1014. Thus it is to reckon one's chickens before they are hatched: and thus the Transalleganian dream vanishes. You have been deceived. A warm heart, a lively imagination, and I know not what caprice about republicanism, have led you into sanguine expectations and wrong conclusions. Come, now! Confess it like yourself; that is, like a man of sense and spirit: like an honest and fair-dealing John Bull. To err belongs to all men, great as well as little; but, to be ashamed to confess error, belongs only to the latter.

1015. Great as is my confidence in your candour, I can, however, hardly hope wholly to escape your anger for having so decidedly condemned your publications; but, I do hope, that you will not be so unjust as to impute my conduct to any base self-interested motive. I have no private interest, I can have no such interest in endeavouring to check the mad torrent towards the West. I *own* nothing in these States, and never shall; and whether English Farmers push on into misery and ruin, or stop here in happiness and prosperity, to me, as far as private interest goes, it must be the same. As to the difference in our feelings and notions about *country*, about *allegiance*, and about *forms of government*, this may exist without any, even the smallest degree of personal dislike. I was no hypocrite in England; I had no views farther than those which I professed. I wanted nothing for myself but the fruit of my own industry and talent, and I wished nothing for my country but its liberties and laws, which say, that the people shall be *fairly represented.* England has been very happy and *free;* her greatness and renown have been surpassed by those of no nation in the world; her wise, just, and merciful laws form the basis of that freedom which we here enjoy, she has been fertile beyond all rivalship in men of learning and men devoted to the cause of freedom and humanity; her people, though proud and domineering, yield to no people in the world in frankness, good faith, sincerity, and benevolence: and I cannot but know, that this state of things has existed, and that this people has been formed, under a government of king, lords, and commons. Having this powerful argument of experience before me, and seeing no reason why the thing should be otherwise, I have never wished for republican government in England; though, rather than that the present tyrannical oligarchy should continue to trample on king and people, I would gladly see the whole fabric torn to atoms, and trust to chance for something better, being sure that nothing could be worse. But, if I am not a republican; if I think my duty towards England indefeasible; if I think that it becomes me to abstain from any act which shall seem to say I abandon her, and especially in this her hour of distress and oppression; and, if, in all these points, I differ from you, I trust that to this difference no part of the above strictures will be imputed, but that the motive will be fairly inferred from the act, and not the act imputed unfairly to any motive. I am, my dear Sir, with great respect for your talents as well as character,

Your most obedient
And most humble servant,

Wm. COBBETT.

TO

MORRIS BIRKBECK, Esq.

OF ENGLISH PRAIRIE, ILLINOIS TERRITORY.

LETTER II.

North Hempstead, Long Island,
15th Dec. 1818.

My dear Sir,

1016. Being, when I wrote my former Letter to you, in great haste to conclude, in order that my son William might take it to England with him, I left unnoticed many things, which I had observed in your " *Letters from the Illinois;*" and which things merited pointed notice. Some of these I will notice; for, I wish to discharge all my duties towards my countrymen faithfully; and, I know of no duty more sacred, than that of warning them against pecuniary ruin and mental misery.

1017. It has always been evident to me, that the Western Countries were not the countries for *English* farmers to settle in: no, nor for American farmers, unless under peculiar circumstances. The settlers, who have gone from the New England States, have, in general, been *able men* with families of *stout sons.* The contracted farm in New England sells for money enough to buy the land for five or six farms in the West. These farms are *made* by the *labour of the owners.* They *hire nobody.* They live *any how* for a while. I will engage that the labour performed by one stout New England family in *one year*, would cost an English farmer *a thousand pounds in wages.* You will say, why cannot the English labour as hard as the Yankees? But, mind, I talk of a *family* of Yankee *sons;* and, besides, I have no scruple to say, that one of these will do as much work in the *clearing* and *fencing* of a farm, and in the *erection of buildings*, as *four or five English* of the same age and size. Yet, have many of the New England farmers *returned.* Even *they* have had cause to repent of their folly. What hope is there, then, that English farmers will succeed?

1018. It so happens, that *I have seen* new settlements formed. I have seen lands cleared. I have seen crowds of people coming and squatting down in woods or little islands, and by the sides of rivers. I have seen the log-hut raised; the bark covering put on; I have heard the bold language of the adventurers; and I have witnessed their subsequent miseries. They were just as *free* as you are; for, they, like you, saw no signs of the existence of any government, good or bad.

1019. New settlements, particularly at so great a distance from all the conveniences and sweeteners of life, must be begun by people who *labour for themselves.* Money is, in such a case, almost useless. It is impossible to believe, that, after your statement about your intended *hundred acres of Indian corn*, you would not have had it, or, at least, a part of it, if you *could;* that is to say, if *money* would have got it. Yet you had not a single square rod. Mr. HULME, (See Journal, 28th July) says, in the way of *reason* for your having *no crops* this year, that you could *purchase* with *more economy than you could grow!* Indeed! what; would the Indian Corn have *cost, then, more than the price of the Corn?* Untoward observation; but *perfectly true*, I am convinced. There is, it is my opinion, nobody that can raise Indian Corn or Grain at so great a distance from a market to any profit at all with *hired labour.* Nay, this is too plain a case to be matter of *opinion.* I may safely assume it as an indisputable fact. For, it being notorious, that labour is as high priced with you as with us, and your statement shewing that Corn is not much more than *one third* of our price, how monstrous, if you gain at all, must be the Consumers' gains here! The *rent* of the land here is a mere trifle more than it must be there, for the cultivated part must pay rent for the uncultivated part. The *labour*, indeed, as all the world knows, is every thing. All the other expences are not worth speaking of. What, then, must be the gains of the Long Island farmer, who sells his corn at a dollar a bushel, if you, with labour at the Long Island price, can *gain* by selling Corn at the rate of *five bushels for two dollars!* If yours be a *fine country* for English farmers to migrate to, *what must this be?* You want *no manure.* This cannot last long; and, accordingly, I see, that you mean *to dung for wheat after the second crop of Corn.* This is another of the romantic stories exposed. In Letter IV you relate the romance of *manure being useless;* but, in Letter X, you tell us, that you propose to *use* it. Land bearing crops without a manure, or, with new-culture and constant ploughing, is a romance. This I told you in London; and this you have found to be true.

1020. It is of little consequence what wild schemes are formed and executed by men who have property enough to *carry them back;* but, to invite men to go to the Illinois with *a few score of pounds* in their pockets, and to tell them, that they can become *farmers* with those pounds, appears to me to admit of no other apology than an unequivocal acknowledgment, that the inviter is *mad.* Yet your *fifteenth* Letter from the Illinois really contains such an invitation. This letter is manifestly addressed to an *imaginary* person. It is clear that the correspondent is a *feigned*, or *supposed*, being. The letter is, I am sorry to say, I think, a mere trap to catch poor creatures with a few pounds in their pockets. I will here take the liberty to insert the whole of this letter; and will then endeavour to show the misery which it is calculated to produce, not only amongst English people, but amongst Americans who may chance to read it, and who are now living happily in the Atlantic States. The letter is dated, 24th of February, 1818, and the following are its words:

1021. "Dear Sir,—When a man gives advice "to his friends, on affairs of great importance "to their interest, he takes on himself a load of "responsibility, from which I have always "shrunk, and generally withdrawn. My *ex-"ample* is very much at their service, either for "imitation or warning, as the case may be. I "must, however, in writing to *you*, step a little "over this line of caution, having more than "once been instrumental in helping you, not "*out* of your difficulties, but from one scene of "perplexity to another; I cannot help advising "you to make an effort more, and extricate "yourself and family completely, by removing "into this country.—When I last saw you, "twelve months ago, I did not think favourably "of your prospects: if things have turned out "better, I shall be rejoiced to hear it, and you "will not need the advice I am preparing for "you. But, if vexation and disappointments "have assailed you, as I feared, and you can "honourably make your escape, with the "means of transmitting yourself hither, and "one hundred pounds sterling to spare—don't "hesitate. In six months after I shall have "welcomed you, barring accidents, you shall "discover that you are become *rich*, for you "shall feel that you are independent: and I "think that will be the most delightful sensa-"tion you ever experienced; for, you will re-"ceive it multiplied, as it were, by the number "of your family as your troubles now are. It "is not, however, a sort of independence that "will excuse you from labour, or afford you "many luxuries, that is, costly luxuries. I "will state to you what I have learned, from a "good deal of observation and inquiry, and a "little experience; then you will form your "own judgment. In the first place, the voyage. "That will cost you, to Baltimore or Philadel-

" phia, provided you take it, as no doubt you
" would, in the cheapest way, twelve guineas
" each, for a birth, fire, and water, for yourself
" and wife, and half price, or less, for your
" children, besides provisions, which you will
" furnish. Then the journey. Over the moun-
" tains to Pittsburgh, down the Ohio to Shaw-
" nee Town, and from thence to our settle-
" ment, fifty miles north, will amount to five
" pounds sterling per head.—If you arrive here
" as early as May, or even June, another five
" pounds per head will carry you on to that
" point, where you may take your leave of de-
" pendence on any thing earthly but your own
" exertions.—At this time I suppose you to have
" remaining one hundred pounds (borrowed
" probably from English friends, who rely on
" your integrity, and who may have directed
" the interest to be paid to me on their behalf,
" and the principal in due season.)—We will
" now, if you please, turn it into dollars, and
" consider how it may be disposed of. A
" hundred pounds sterling will go a great way
" in dollars. With eighty dollars you will en-
" ter a quarter section of land; that is, you
" will purchase at the land-office one hundred
" and sixty acres, and pay one-fourth of the
" purchase money, and looking to the land to
" reward your pains with the means of dis-

" charging the other three-fourths as they be-
" come due, in two, three, and four years.—
" You will build a house with fifty dollars;
" and you will find it extremely comfortable
" and convenient, as it will be really and truly
" yours. Two horses will cost, with harness
" and plough, one hundred.—Cows, and hogs,
" and seed corn, and fencing, with other ex-
" penses, will require the remaining two hund-
" red and ten dollars.—This beginning, humble
" as it appears, is affluence and splendour, com-
" pared with the original outfit of settlers in
" general. Yet no man remains in poverty, who
" possesses even moderate industry and eco-
" nomy, and especially of *time*.—You would of
" course bring with you your sea-bedding and
" store of blankets, for you will need them on
" the Ohio; and you should leave England
" with a good stock of wearing apparel. Your
" luggage must be composed of light articles,
" on account of the costly land-carriage from
" the Eastern port to Pittsburgh, which will
" be from seven to ten dollars per 100 lbs.
" nearly sixpence sterling per pound. A few
" simple medicines of good quality are indis-
" pensable, such as calomel, bark in powder,
" castor oil, calcined magnesia, laudanum;
" they may be of the greatest importance on
" the voyage and journey, as well as after

" your arrival.—Change of climate and situa-
" tion will produce temporary indisposition,
" but with prompt and judicious treatment,
" which is happily of the most simple kind, the
" complaints to which new comers are liable are
" seldom dangerous or difficult to overcome,
" provided due regard has been had to salubrity
" in the choice of their settlement, and to diet
" and accommodation after their arrival.

" With best regards, I remain, &c."

1022. Now, my dear sir, your mode of address, in this letter, clearly shews that you have in your eye a person above the level of common labourers. The words "*Dear Sir*" indicate that you are speaking to a *friend*, or, at least, to an *intimate acquaintance*; of course to a person, who has not been brought up in the habits of *hard labour*. And such a person it is, whom you advise and press to come to the Illinois with a *hundred pounds in his pocket* to become a *farmer!*

1023. I will pass over the expences previous to this unfortunate man and his family's arriving at the Prairies, though those expences will be *double* the amount that you state them at. But he arrives with 450 dollars in his pocket. Of these he is to pay down 80 for his land, leaving three times that sum to be paid afterwards.

He has 370 left. And now what is he to do? He arrives *in May*. So that this family has to cross the sea in *winter*, and the land in *spring*. There they are, however, and now what are they to *do?* They are to have built for 50 dollars *a house* "EXTREMELY COM-"FORTABLE AND CONVENIENT:"—the very words that you use in describing the farmer's house, that was to cost, with out-buildings, 1500 dollars! However, you have described your own *cabin*, whence we may gather the meaning which you attach to the word *comfortable*. "This cabin is built of round "straight logs, about a foot in diameter, laying "upon each other, and notched in at the cor-"ners, forming a room eighteen feet long by "sixteen; the intervals between the logs "'chunked,' that is, filled in with slips of "wood; and 'mudded,' that is, daubed with a "plaister of mud; a spacious *chimney, built* "*also of logs*, stands like a bastion at one end; "the roof is well covered with four hundred "'clap boards' of cleft oak, very much like the "pales used in England for fencing parks. A "hole is cut through the side called, very pro-"perly, the '*through*,' for which there is a "'shutter,' made also of cleft oak, and hung on "wooden hinges. All this has been executed "by contract, and well executed, for *twenty*

" *dollars.* I have since added *ten dollars* to the
" cost, for the *luxury* of a *floor* and *ceiling* of
" sawn boards, and it is now a *comfortable* ha-
" bitation."

1024. In plain words, this is a *log-hut*, such as the free negroes live in about here, and a hole it is, fit only for dogs, or hogs, or cattle. Worse it is than the negro huts; for they have a bit of *glass;* but here is none. This miserable hole, black with smoke as it always must be, and without any window, costs, however, 30 dollars. And yet this English acquaintance of yours is to have "a *house extremely comfortable* " *and convenient for fifty dollars.*" Perhaps his 50 dollars might get him a hut, or hole, a few feet longer and divided into two dens. So that here is to be *cooking, washing, eating*, and *sleeping* all in the same " *extremely* convenient " and comfortable" hole! And yet, my dear Sir, you find fault of the want of *cleanliness* in the *Americans!* You have not seen " *the Ame-* " *ricans.*" You have not seen the nice, clean, neat houses of the farmers in this Island, in New England, in the Quaker counties of Pennsylvania. You have seen nothing but the smoke-dried Ultra-montanians; and your project seems to be to make the deluded English who may follow you rivals in the attainment of the tawny colour. What is this family to do in their 50 dollar den? Suppose one or more of them *sick!* How are the rest to sleep by night or to eat by day?

1025. However, here they are, in this miserable place, with the *ship-bedding*, and without even a bedstead, and with 130 dollars gone in land and house. *Two horses and harness and plough* are to cost 100 dollars! These, like the *hinges* of the door, are all to be of *wood* I suppose; for as to flesh and blood and bones in the form of two horses for 100 dollars is impossible, to say nothing about the plough and harness, which would cost 20 dollars of the money. Perhaps, however, you may mean some of those horses, ploughs and sets of harness, which, at the time when you wrote this letter, you had *all ready waiting for the spring to put in your hundred acres of corn* that was *never put in at all!* However, let this pass too. Then there are 220 dollars left, and these are to provide *cows, hogs, seed, corn, fencing*, and other *expences.* Next come two cows (poor ones) 24 dollars; hogs, 15 dollars; seed corn, 5 dollars; fencing, suppose 20 acres only, in four plots, the stuff brought from the woods nearest adjoining. Here are 360 rods of fencing, and, if it be done so as to keep out a pig, and to keep in a pig, or a horse or cow, for less than half a dollar a rod, I will suffer myself to be made into smoked meat in the extremely comfortable house. Thus, then, here are 213 out of the 220 dollars, and this happy settler has *seven* whole dollars left for all "*other expences*; amongst which are the cost of cooking utensils, plates, knives and forks, tables, and stools; for, as to *table-cloths* and *chairs*, those are luxuries unbecoming "simple "republicans." But, there must be a *pot* to boil in; or, is that too much? May these republicans have a washing tub? Perhaps, indeed, it will become unnecessary in a short time; for, the lice will have eaten up the linen; and, besides, perhaps real *independence* means stark-nakedness. But, at any rate, the hogs must have a *trough;* or, are they to eat at the same board with the family? Talking of *eating* puts me in mind of a great article; for what are the family to *eat* during *the year and more* before their land can produce? For even if they arrive in *May*, they can have *no crop that year.* Why, they must graze with the cows in the Prairies, or snuggle with the hogs in the woods. An *oven!* Childish effeminacy! Oh! unleavened bread for your life. *Bread*, did I say? Where is the "independent" family to get bread? Oh! no! Grass and Acorns and Roots; and, God be praised, you have plenty of water in your wells, though, perhaps, the family, with all their "*independence*," must be compelled to depend on your leave to get it, and fetch it half a mile into the bargain.

1026. To talk seriously upon such a subject is impossible, without dealing in terms of reprobation, which it would give me great pain to employ when speaking of any act of yours. Indeed such a family will be *free;* but, the Indians are free, and so are the gypsies in England. And I most solemnly declare, that I would sooner live the life of a gypsy in England, than be a settler, with less than five thousand pounds, in the Illinois; and, if I had the five thousand pounds, and was resolved to exchange England for America, what in the name of common sense, should induce me to go into a wild country, when I could buy a good farm of 200 acres, with fine orchard and good house and out-buildings, and stock it completely, and make it rich as a garden, within twenty miles of a great sea-port, affording me a ready market and a high price for every article of my produce?

1027. You have, *by this time*, seen more than you had seen, when you wrote your "Letters "from the Illinois." You would not, I am convinced, write such letters *now*. But, lest you should not do it, it is right that somebody should counteract their delusive effects; and and this I endeavour to do as much for the sake of this country as for that of my own country-men. For a good while I remained silent, hoping that few people would be deluded; but when I heard, that an old friend, and brother sportsman; a sensible, honest, frank, and friendly man, in *Oxfordshire*, whom I will not name, had been seized with the Illinois madness, and when I recollected, that he was one of those, who came to *visit me in prison*, I could no longer hold my tongue; for, if a man like him; a man of his sound understanding, could be carried away by your representations, to what an extent must the rage have gone!

1028. Mr. HULME visited you with the most friendly feelings. He agrees with you perfectly as to notions about forms of government. He *wished* to give a good account of your proceedings. His account is favourable; but, his *facts*, which I am sure are true, let out what I could not have known for certainty from any other quarter. However, I do not care a farthing for the *degrees* of goodness or of badness; I say all new countries are *all badness* for *English farmers.* I say, that *their* place is near the great cities on the coast; and that every step they go beyond forty miles from those cities is a step *too far*. They want freedom: they have it here. They want good land, good roads, good markets: they have them all here. What should they run rambling about a nation-making for? What have they to do about extending dominion and "taming the wilderness?" If they speculate upon becoming founders of republics, they will, indeed, do well to get out of the reach of rivals. If they have a thirst for power, they will naturally seek to be amongst the least informed part of mankind. But, if they only want to keep their property and live well, they will take up their abode on this side of the mountains at least.

1029. The *grand ideas* about the *extension of the empire* of the United States are of very questionable soundness: and they become more questionable from being echoed by the *Edinburgh Reviewers*, a set of the meanest politicians that ever touched pen and paper. Upon any great question, they never have been right, even by *accident*, which is very hard! The *rapid* extension of settlements to the West of the mountains is, in my opinion, by no means favourable to the duration of the present happy Union. The conquest of Canada would have been as dangerous; but not more dangerous. A nation is never so strong and so safe as when its extreme points feel for each other as acutely as each feels for itself; and this never can be when all are not equally exposed to every danger; and especially when all the parts have not the *same interests.* In case of a war with England, what would become of your market down the Mississippi? That is your sole market. That way your produce must go; or you must dress yourself in skins and tear your food to bits with your hands. Yet that way your produce could not go, unless this nation were to keep up a Navy equal to that of England. Defend the country against invaders I know the people always will; but, I am not sure, that they will like internal taxes sufficient to rear and support a navy sufficient to clear the gulph of Mexico of English squadrons. In short, it is my decided opinion, that the sooner the banks of the Ohio, the Wabash, and the Mississippi are pretty thickly settled, the sooner the *Union* will be placed in jeopardy. If a war were to break out with England, even in a few years, the lands of which the Mississippi is the outlet, would lose a great part of their value. Who does not see in this fact a great cause of *disunion?* On this side the mountains, there are twelve hundred miles of coast to blockade; but you, gentlemen Prairie owners, are like a rat that has but one hole to go out and to come in at. You express your deep-rooted attachment to your adopted country, and I am sure you are sincere; but, still I may be allowed to doubt, whether you would cheerfully wear bear-skins, and gnaw your meat off the bones for the sake of any commercial right that the nation might go to war about. I know that you would not *starve*; for coffee and tea are not necessary to man's existence; but, you would like to sell your flour and pork, and would be very apt to discover reasons against a war that would prevent you from selling them. You appear to think it very wicked in the Atlantic People to feel little eagerness in promoting the increase of population to the Westward; but, you see, that, in this want of such eagerness, they may be actuated by a real love for their country. For my part, I think it would have been good policy in the Congress not to dispose of the Western Lands at all; and I am sure it would have been an act of real charity.

1030. Having now performed what I deemed my duty towards my countrymen, and towards this country too, I will conclude my letter with a few observations, relative to *mills*, which may be of use to *you;* for, I know, that you will *go on;* and, indeed, I most sincerely wish you all the success that you can wish yourself, without doing harm to others.

1031. You have *no mill streams* near you; and you are about to erect a *wind-mill.* Man is naturally prone to call to his aid whatever will save his bones labour. The *water*, the *wind*, the *fire:* any thing that will *help* him. Cattle of some sort or other were, for a long while, his great resource. But, of late, water-powers, wind-powers, fire-powers. And, indeed, wondrous things have been performed by machines of this kind. The water and the wind do *not eat*, and require no grooming. But, it sometimes happens, that, when all things are considered, we resort to these grand powers without any necessity for it; and that we forget how easily we could do the thing we want done, with our own hands. The story, in Peregrine Pickle, about the Mechanic, who had invented a *water machine to cut off the head of a cabbage*, hardly surpassed the reality in the case of the machine, brought out in England, some years ago, *for reaping wheat;* nor is it much less ridiculous to see people going many miles with grist to a mill, which grist they might so easily grind at home. The hand-mills, used in England, would be invaluable with you, for a while, at least.

1032. But, it is of a mill of more general utility, that I am now about to speak to you; and, I seriously recommend it to your consideration, as well as to other persons similarly situated.

1033. At Botley I lived surrounded by water-mills and wind-mills. There were eight or ten within five miles of me, and one at two hundred yards from my house. Still I thought, that it was a brutal sort of thing to be obliged to send twice to a mill, with all the uncertainties of the business, in order to have a sack of wheat or of barley ground. I sent for a mill-wright, and, after making all the calculations, I resolved to have a mill in my farm yard, to grind for myself, and to sell my wheat in the shape of flour. I had the mill erected in a pretty little barn, well floored with oak, and standing upon stones with caps: so that no rats or mice could annoy me. The mill was to be moved by *horses*, for which, to shelter them from the wet, I had a shed with a circular roof erected on the outside of the barn. Under this roof, as well as I recollect, there was a large wheel, which the horses turned, and a bar, going from that wheel, passed through into the barn, and there it put the whole machinery in motion.

1034. I have no skill in mechanics. I do not, and did not, know one thing from another by its name. All I looked to was the *effect;* and this was complete. I had excellent flour. All my meal was ground at home. I was never bothered with *sending to the mill.* My ears were never after dinned with complaints about *bad flour* and *heavy bread.* It was the prettiest, most convenient, and most valuable thing I had upon my farm. It was, I think, put up in 1816, and this was one of the pleasures, from which the Borough-villains (God confound them!) drove me in 1817. I think it cost me about *a hundred pounds.* I forget, whether I had *sold* any flour from it to the Bakers. But, independent of that, it was very valuable. I think we ground and dressed about *forty bushels* of wheat in a day; and, we used to work at it on *wet days*, and when we could not work in the fields. We never were stopped by want of wind or water. The horses were always ready; and *I know*, that our grinding was done at *one half* the expence at which it was done by the millers.

1035. The farmers and millers used to say, that I *saved nothing* by my mill. Indeed, *gain* was not my object, except in *convenience.* I hated the *sudden* calls for going to the mill. They produced *irregularity;* and, besides, the millers were not *more honest* than other people, Their mills contained all sorts of grain; and, in their confusion, we sometimes got *bad flour* from *good wheat*; an accident that never happened to us after we got our own mill. But, as to the *gain*, I have just received a letter from my son, informing me, that the gentleman, a farmer born and bred, who rents my farm in my absence, *sells no wheat;* that he *grinds all;* that he *sells flour* all round the country; and that this flour is *preferred before that of the millers.* I was quite delighted to hear this news of my little mill. It awakened many recollections; and I immediately thought of communicating the facts to the public, and particularly *to you.*

1036. You will observe, that my farm is situated in *the midst of mills.* So that, you may be sure, that the thing answers, or it would not be carried on. If it were not attended with *gain,* it would not be put in motion. I was convinced, that any man might grind *cheaper* with a horse-mill than with a water or windmill, and now the fact is proved. For, observe, the mill costs nothing for *scite;* it occupies a very small space; it is independent of wind and water; no floods or gales can affect it.

1037. Now, then, if such a mill be preferable to wind or water-mills in a place where both abound, how useful must it be in a situation like yours? Such a mill would amply supply about three hundred families, if kept constantly at work. And then, it is so much more convenient than a windmill. A windmill is necessarily a most unhandy thing. The grain has to be hauled up and the flour let down. The building is a place of *no capacity;* and, there is great danger attending the management of it.

My project is merely a neat, close barn, standing upon stones that rats and mice cannot creep up. The waggon comes to the door, the sacks are handed in and out; and every thing is so convenient and easily performed, that it is a pleasure to behold it.

1038. About the construction of the mill I know nothing. I know only the *effect*, and that it is worked by horses, in the manner that I have described. I had *no Miller*. My Bailiff, whom I had made a Bailiff out of a Carpenter, I turned into a Miller; or, rather, I made him look after the thing. Any of the men, however, could do the millering very well. Any of them could make *better flour* than the water and wind-millers used to make for us. So that there is no *mystery* in the matter.

1039. This country abounds in excellent millwrights. The best, I dare say, in the world; and, if I were settled here as a farmer in a large way, I would soon have a little mill, and send away my produce in flour instead of wheat. If a farmer has to send *frequently* to the mill, (and that he must do, if he have a great quantity of stock and a large family,) the very expence of *sending* will pay for a mill in two or three years.

1040. I shall be glad if this piece of information should be of use to any body, and particularly if it should be of any use in the Prairies; for, God knows, you will have plague enough without *sending to mill,* which is, of itself, no small plague even in a Christian country. About the same strength that turns a threshing machine, turned my mill. I can give no information about the construction. I know there was a *hopper* and *stones,* and that the thing made a *clinking* noise like the water-mills. I know that the whole affair occupied but a small space. My barn was about forty feet long and eighteen feet wide, and the mill stood at one end of it. The man who made it for me, and with whom I made a bargain in writing, wanted me to agree to a *specification* of the *thing;* but I declined having any thing to do with *cogs* and *wheels,* and persisted in stipulating for *effects.* And these were, that with a certain force of horses, it was to make so much fine flour in so long a time; and this bargain he very faithfully fulfilled. The price was I think seventy pounds, and the putting up and altogether made the amount about a hundred pounds. There were no *heavy timbers* in any part of the thing. There was not a bit of wood, in any part of the construction, so big as my thigh. The whole thing might have been carried away, all at once, very conveniently, in one of my waggons.

1041. There is another thing, which I beg leave to recommend to your attention; and that is, the use of the *Broom-Corn Stalks* as *thatch*. The coverings of barns and other out-houses with *shingles* makes them fiery hot in summer, so that it is dangerous to be at work in making mows near them in very hot weather. The heat they cause in the upper parts of houses, though there be a ceiling under them is intolerable. In the *very hot weather* I always bring my bed down to the ground-floor. Thatch is cool. Cool in summer and warm in winter. Its inconveniences are *danger from fire* and *want of durability*. The former is no great deal greater than that of shingles. The latter may be wholly removed by the use of the *Broom-Corn Stalks*. In England a good thatch of wheat-straw will last twelve or fifteen years. If this straw be *reeded*, as they do it in the counties of Dorset and Devon, it will last thirty years; and it is very beautiful. The little town of CHARMOUTH, which is all thatched, is one of the prettiest places I ever saw. What beautiful thatching might be made in *this* country, where the straw is so sound and so clean! A Dorsetshire thatcher might, upon this very island, make himself a decent fortune in a few years. They do cover barns with straw here sometimes; but how one of our thatchers would laugh at the work! Let me digress here, for a moment, to ask you if you have got a *sow-spayer?* We have no such man here. What a loss arises from this! What a plague it is. We cannot keep a whole farrow of pigs, unless we breed from all the sows! They go away: they plague us to death. Many a man in England, now as poor as an owlet, would (if he kept from the infernal drink) become *rich* here in a short time. These sow-gelders, as they call them, swarm in England. Any clown of a fellow follows this calling, which is hardly two degrees above rat-catching and mole-catching: and yet there is no such person here, where swine are so numerous, and where so many millions are fatted for exportation! It is very strange.

1042. To return to the thatching: Straw is not so durable as one could wish: besides, in very high winds, it is liable, if not *reeded,* to be ruffed a good deal; and the reeding, which is almost like counting the straws one by one, is expensive. In England we sometimes thatch with *reeds,* which in Hampshire, are called *spear.* This is an aquatic plant. It grows in *the water*, and will grow no where else. When stout it is of the thickness of a small cane at the bottom, and is about four or five feet long. I have seen a thatch of it, which, with a little patching, had lasted upwards of *fifty years.* In gentlemen's gardens, there are sometimes hedges or screens made of these reeds. They last, if well put up, half a century, and are singularly neat, while they parry the wind much better than paling or walls, because there is no eddy proceeding from their repulsion. They are generally put round those parts of the garden where the hot-beds are.

1043. Now, the Broom-Corn far surpasses the reeds in all respects. I intend, *in my Book on Gardening*, to give a full account of the applicability of this plant to *garden-uses* both here and in England; for, as to the *reeds*, they can seldom be had, and a screen of them comes, in most parts of England, to *more money than a paling of oak*. But, the Broom-Corn! What an useful thing! What quantities upon an acre of land! *Ten feet high*, and *more durable than reeds!* The *seed-stems*, with a bit of the stem of the plant, make the brooms. These, I hear, are now *sent to England*. I have often talked of it in England as a good traffic. We here sweep *stables* and *streets* with what the English sweep their *carpets* with! You can buy as good a broom at New York for *eight pence* sterling as you can buy in London for *five shillings sterling*, and the freight cannot exceed two-pence or three-pence, if sent without handles. I bought a clothes-brush, an

English clothes-brush, the other day for *three shillings sterling*. It was made of *a farthing's worth of alder wood* and of *half a farthing's worth of Broom-Corn.* An excellent brush. Better than bristles. I have Broom-Corn and Seed-Stems enough to make fifty thousand such brushes. I really think I shall send it to England. It is now lying about my barn, and the chickens are living upon the seeds. This plant demands *greater heat* even than the Indian Corn. It would hardly *ripen its seed* in England. Indeed it would not. But, if well managed, it would produce a prodigious crop of materials for reed-hedges and thatch. It is of a substance (I mean the main stalk) between that of a *cane* and that of a *reed.* It has joints precisely like those of the canes, which you may have seen the Boroughmongers' sons and footmen strut about with, called *bamboos.* The *seed-stalks,* which make the brooms and brushes, might not get so *mature* in England as to be *so good* as they are here for those uses: but, I have no doubt, that, in any of the warm lands in Surry, or Kent, or Hampshire, a man might raise upon an acre a crop worth *several hundred pounds.* The *very stout* stalks, if properly harvested and applied, would last nearly as long as the best *hurdle rods.* What beautiful screens they would make in gardens and pleasure grounds! Ten feet long, and straight as a gun stick! I shall send some of the seed to England this year, and cause a trial to be made; and I will, in my Gardening Book, give full instructions for the cultivation. Of this book, which will be published soon, I would, if you lived in this world, send you a copy. These are the *best uses* of maritime intercourse: the interchange of plants, animals, and improvements of all sorts. I am doing my best to repay this country for the protection which it has given me against our *indemnified* tyrants. "Cob-" bett's pigs and Swedish Turnips" will be talked of long after the bones of Ellenborough, Gibbs, Sidmouth, Castlereagh and Jenkinson will be rotten, and their names forgotten, or only remembered when my "trash" shall.

1044. This is a rambling sort of Letter. I now come back to the Broom-Corn for thatch. Sow it in *rows* about five feet asunder; or, rather, on *ridges*, a *foot wide at the top*, with an interval of *five feet;* let the plants stand all over this foot wide, at about three inches apart, or less. Keep the plants clear of weeds by a couple of weedings, and *plough well* between the ridges three or four times during the summer. This will make the plants grow *tall,* while their closeness to each other will make them *small in thickness* of stem or stalk. It will bring them to about the *thickness* of fine large reeds in England, and to about twice the length; and, I will engage, that a large barn may be covered, by a good thatcher, with the stalks, in *two days*, and that the covering shall last for fifty years. Only think of the price of *shingles* and *nails!* Only think of the cost of *tiles* in England! Only think of the expence of drawing or of reeding straw in England! Only think of *going into the water* to collect reeds in England, even where they are to be had at all, which is in a very few places! The very first thing that I would do, if I were to settle in a place where I had buildings to erect, would be to sow some Broom-Corn; that is to say, sow some *roofs*. What a fine thing this would be upon the farms in England! What a convenient thing for the cottagers! Thatch for their pretty little houses, for their styes, for their fuel-house, their cow-shed; and brooms into the bargain; for, though the *seed* would not *ripen*, and though the broom-part would not be of the *best* quality, it would be a thousand times better than *heath*. The seed might be sent from this country, and, though the Borough-villains would *tax* it, as their rapacious system does EVEN THE SEEDS OF TREES; yet, a small quantity of seed would suffice.

1045. As an *ornamental* plant nothing equals this. The Indian Corn is far inferior to it in this respect. Planted by the side of walks in gardens, what beautiful avenues it would make for the summer! I have seen the plants *eighteen feet and a half high.* I always wanted to get some seed in England; but, I never could. My friends thought it too *childish* and *whimsical* a thing to attend to. If the plant should so far come to perfection in England as to yield the broom-materials, it will be a great thing; and, if it fall short of that, it will certainly surpass *reeds* for thatching and screening purposes, for sheep-yards, and for various other uses. However, I have no doubt of its producing *brooms;* for, the Indian Corn, though only certain sorts of it will ripen its seed even in Hampshire, will always come into *bloom*, and, in the Broom-Corn, it is the little stalks, or branches, out of which the flower comes, that makes the broom. If the plant succeed thus far in England, you may be sure that the Borough-villains will *tax* the brooms, until their system be blown to atoms; and, I should not wonder if they were to make the broom, like hops, an *article of excise*, and send their spies into people's fields and gardens to see that the revenue was not "*defrauded.*" Precious villains! They stand between the people and all the gifts of nature! But this cannot *last.*

1046. I am happy to tell you, that *Ellenborough* and *Gibbs* have *retired!* Ill health is the *pretence.* I never yet knew ill health induce such fellows to loosen their grasp of the public purse. But, be it so: then I feel pleasure on that account. To all the other pangs of body and mind let them add that of knowing, that William Cobbett, whom they thought they had put down for ever, if not killed, lives to rejoice at their pains and their death, to trample on their graves, and to hand down their names for the just judgment of posterity. What! are these feelings *wrong?* Are they *sinful?* What defence have we, then, against tyranny? If the oppressor be not to experience the resentment of the oppressed, let us at once acknowledge the divine right of tyranny; for, what has tyranny else to fear? Who has it to fear, but those whom it has injured? It is the aggregate of individual injury that makes up national injury: it is the aggregate of individual resentment that makes up national resentment. National resentment is absolutely necessary to the producing of redress for oppression; and, therefore, to say that individual resentment is wrong, is to say, that there ought to be no redress for oppression: it is, in short, to pass a sentence of never-ending slavery on all mankind. Some Local Militia men; young fellows who had been *compelled* to become soldiers, and who had no knowledge of military discipline; who had, by the Act of Parliament, been promised *a guinea each* before they marched; who had refused to march *because the guinea had not been wholly paid them;* some of these young men, these mere boys, had, for this mutiny, as it was called, been *flogged* at Ely in Cambridgeshire, under a guard of *German bayonets* and *sabres.* At this I expressed *my indignation in the strongest terms;* and, for doing this, I was put for *two years* into a jail along with men convicted of *unnatural crimes, robbery*, and under *charge of murder*, and where ASTLET was, who was *under sentence of death.* To this was added *a fine of a thousand pounds sterling;* and, when the two years should expire, *bonds for the peace and good behaviour* for *seven years!* The seven years are not yet expired. I will endeavour to be of "*good be-*"*haviour*" for the short space that is to come; and, I am sure, I have behaved well for the *past;* for never were seven years of such *efficient exertion* seen in the life of any individual.

1047. The tyrants are hard pushed now. The *Bank Notes* are their only ground to stand on; and that ground *will be moved from under them in a little time.* Strange changes since you left England, short as the time has been!

I am fully of opinion, that my *four years* which I gave the system at my coming away, will see the end of it. There can be *no more war* carried on by them. I see they have had Baring, of Loan-notoriety at the Holy Alliance-Congress. He has been stipulating for a supply of paper-money. They should have got my consent to let the paper-money remain; for, *I can destroy it whenever I please.* All sorts of projects are on foot. "*Inimitable Notes;*" paying in specie *by weight of metal.* Oh! the wondrous fools! A sudden blow-up; or, a blow-up somewhat slow, by ruin and starvation; one of these *must come;* unless they speedily *reduce the interest of the Debt;* and even that will not save the seat-dealers.

1048. In the meanwhile let us enjoy ourselves here amongst this kind and hospitable people; but, let us never forget, that England is our country, and that her freedom and renown ought to be as dear to us as the blood in our veins. God bless you, and give you health and happiness.

WM. COBBETT.

POSTSCRIPT.

RUTA BAGA; OR, SWEDISH TURNIP.

To the Editor of the New York Evening Post.

Hyde Park, Long-Island,
3d Jan. 1819.

SIR,

1049. MY publications of last year, on the *amount of the crops* of Ruta Baga, were, by many persons, considered *romantic* ; or, at best, a good deal *strained.* I am happy, therefore, to be able to communicate to the public, through your obliging columns, a letter from an *American farmer* on the subject. You may remember, if you did me the honour to read my Treatise on the cultivation of this root (in Part I. of the Year's Residence), that I carried the amount of my best Botley-crops no higher than *one thousand three hundred* bushels to the acre. The following interesting letter will, I think, convince every one, that I kept, in all my statements, below the mark. *Here* we have *an average* weight of roots of *six pounds and a half.*

1050. I beg Mr. Townsend to accept of my best thanks for his letter, which has given me very great satisfaction, and which will, I am sure, be of great use in promoting the cultivation of this valuable root.

1051. Many gentlemen have written to me with regard to the mode of *preserving* the Ruta Baga. I have, in the SECOND PART of my *Year's Residence*, which will be published at New York, in *a few days*, given a very full account of this matter.

I am, Sir, your most humble
And most obedient servant,
Wm. COBBETT.

New York, Dec. 30, 1818.

Dear Sir,

1052. I take the liberty of sending to you the following experiments upon the culture of your Ruta Baga, made by my uncle, Isaac Townsend, Esq. of Orange county, in this state. The seeds were procured from your stock, and the experiments, I think, will tend to corroborate the sentiments which you have so laudably and so successfully inculcated on the subject of this interesting article of agriculture.

1053. A piece of strong dry loam ten feet square on the N. E. side of a mountain in Moreau township, Orange county, was thoroughly cleared of stones, and dug up twelve inches deep, on the 10th of June last; it was then covered by a mixture of ten bushels of charcoal dust and twenty bushels of black swamp mould, which was well harrowed in. About the 9th of July it was sown with your Ruta Baga in drills of twenty inches apart, the turnips being ten inches distant from each other. They came up badly and were weeded out on the 10th of August. On the 15th of August a table-spoonful of ashes was put round every turnip, which operation was repeated on the 20th of September. The ground was kept perfectly clean through the whole season. Six seeds of the common turnip were by accident dropped into the patch, and received the same attention as the rest. These common turnips weighed two pounds a piece. The whole yield of the Ruta Baga was three bushels, each turnip weighing from four to eight pounds. The roots penetrated about twelve inches into the ground, although the season was remarkably dry.

1054. A piece of rich, moist, loamy land, containing four square rods, was ploughed twice in June, and the seeds of your Ruta Baga sown on the 4th of July *in broad cast*, and kept clean through the season. This patch produced *twenty-five bushels* of turnips, each turnip weighing from four to nine pounds. This, you perceive, is at the enormous rate of 1000 *bushels an acre!*

1055. It is Mr. Townsend's opinion, that on some of the soils of Orange County your Ruta Baga may be made to yield 1500 bushels an acre.

I remain, with much respect,

Your obedient servant,

P. S. TOWNSEND.

William Cobbett, Esq.
Hyde Park, Long Island.

SECOND POSTSCRIPT.

FEARON'S FALSEHOODS.

To the Editor of the National Advocate.

Hyde Park, Jan. 9th, 1819.

SIR,

1056. BEFORE I saw your paper of the day before yesterday, giving some extracts from a book published in England by one Fearon, I had written part of the following article, and had prepared to send it home as part of a Register, of which I send one every week. Your paper enabled me to make an addition to the article; and, in the few words below, I have this day sent the whole off to be published in London. If you think it worth inserting, I beg you to have the goodness to give it a place; and I beg the same favour at the hands of all those editors who may have published Fearon's account of what he calls *his visit* to me.

I am, Sir,

Your most obedient,

And most humble servant,

WM. COBBETT.

1057. There is, I am told, one FEARON, who has gone home and written and published a book, *abusing this country and its people in the grossest manner.* I only hear of it by letter. I hear, also, that he *speaks* of *me* as if he *knew* me. I will tell you how far he knew me: I live at a country house 20 miles from New York. One morning, in the summer of 1817, a young man came into the hall, and introduced himself to me under the name of FEARON. The following I find about him in my journal:—
" A Mr. FEARON came this morning and had
" breakfast with us. Told us an odd story
" about having slept in a black woman's hut
" last night for sixpence, though there are excel-
" lent taverns at every two miles along the road.
" Told us a still odder story about his being an
" envoy from a *host of families* in London, to
" look out for a place of settlement in America;
" but he took special care *not to name* any one
" of those families, though we asked him to do
" it. We took him, at first, for a sort of *spy.*
" William thinks he is a shopkeeper's clerk; I
" think he has been a tailor. I observed that he
" carried his elbow close to his sides, and his
" arms, below the elbow, in a horizontal posi-
" tion. It came out that he had been with
" BUCHANAN, Castlereagh's consul at New
" York; but it is too ridiculous; such a thing

" as this cannot be a spy; he can get access no
" where but to taverns and boarding houses."

1058. This note now stands in my journal or diary of 22d August, 1817. I remember that he asked me some very silly questions about the *prices of land, cattle,* and other things, which I answered very shortly. He asked my *advice* about the families emigrating, and the very words I uttered in answer, were these: " Every thing I can say, in such a case, is to " *discourage* the enterprize. If Englishmen " come here, let them come individually, and " sit down amongst the natives: no other plan " is rational."

1059. What I have heard of this man since, is, that he spent his time, or great part of it, in New York, amongst the idle and dissolute young Englishmen, whose laziness and extravagance had put them in a state to make them uneasy, and to make them unnoticed by respectable people. That country must be *bad,* to be sure, which would not give them ease and *abundance* without *labour* or *economy.*

1060. Now, what can such a man know of America? He has not kept house; he has had no being in any neighbourhood; he has never had any *circle of acquaintances* amongst the people; he has never been a *guest* under any of their roofs; he knows nothing of their manners or their characters; and how can such a man be a judge of the effects of their institutions, civil, political, or religious?

1061. I have no doubt, however, that the *reviews* and *newspapers*, in the pay of the Boroughmongers, will do their best to propagate the falsehoods contained in this man's book. But what would you say of the people of America, if they were to affect to believe what the *French General* said of the people of England? This man, in a book which he published in France, said, that all the English married women *got drunk*, and *swore* like troopers; and that all the young women were strumpets, and that the *greater part of them had bastards before they were married.* Now, if the people of America were to affect to *believe this*, what should *we* say of them? Yet, this is just as *true* as this FEARON's account of the people of America.

1062. As to the facts of this man's *visit to me*, my son William, who is, by this time, in London, can and will vouch for their truth at any time, and, if necessary, to Fearon's face, if Fearon has a face which he dares show.

1063. Since writing the above, the New York papers have brought me a specimen of Mr. FEARON's performance. I shall notice only his account of his *visit to me*. It is in the following words:

1064. "*A Visit to Mr. Cobbett.*---Upon arriving
" at Mr. Cobbett's gate, my feelings, in walking
" along the path which led to the residence of
" this celebrated man are difficult to describe.
" The idea of a person self-banished, leading an
" isolated life in a foreign land; a path rarely
" trod, *fences in ruins, the gate broken, a house*
" *mouldering to decay*, added to much awk-
" wardness of feeling on my part, calling upon
" an entire stranger, produced in my mind feel-
" ings of thoughtfulness and melancholy. I
" would fain almost have returned without en-
" tering the wooden mansion, imagining that its
" possessor would exclaim, 'What intruding
" fellow is here coming to break in upon my
" pursuits?' But these difficulties ceased almost
" with their existence. A female servant (an
" English woman) informed me that her master
" was from home, attending at the county court.
" Her language was natural enough for a per-
" son in her situation; she pressed me to walk
" in, being quite certain *that I was her country-*
" *man;* and she was so *delighted to see an Eng-*
" *lishman*, instead of those *nasty guessing Yan-*
" *kees.* Following my guide through the
" kitchen, (the floor of which, she asserted,
" was *imbedded with two feet of dirt when Mr.*
" *Cobbett came there*—(it had been previously
" in the occupation of *Americans*) I was con-

" ducted to a front parlour, which contained
" but a *single chair* and several trunks of sea-
" clothes. Mr. Cobbett's first question on
" seeing me was, 'Are you an American, sir?'
" then, 'What were my objects in the United
" States? Was I acquainted with the friends
" of liberty in London? How long had I left?'
" &c. He was immediately familiar. I was
" pleasingly disappointed with the general tone
" of his manners. Mr. Cobbett *thinks meanly*
" *of the American people*, but spoke highly of
" the economy of their government.—He does
" not advise persons in respectable circum-
" stances to emigrate, even in the present state
" of England. In his opinion a family who
" can barely live upon their property, will
" more consult their happiness by not removing
" to the United States. He almost *laughs at*
" *Mr. Birkbeck's* settling in the western coun-
" try. This being the first time I had seen this
" well-known character, I viewed him with no
" ordinary degree of interest. A print by Bar-
" tolozzi, executed in 1801, conveys a correct
" outline of his person. His eyes are small,
" and pleasingly good natured. To a French
" gentleman present, he was attentive; with
" his sons, familiar; to his servants, easy; but
" to all, in his tone and manner, resolute and
" determined. He feels no hesitation in prais-

" ing himself, and evidently believes that he is
" eventually destined to be the Atlas of the Bri-
" tish nation. His faculty of relating anec-
" dotes is amusing. Instances when we meet.
" My impressions of Mr. Cobbett are, that
" those who know him would like him, if they
" can be content to submit unconditionally to
" his dictation. 'Obey me, and I will treat you
" kindly; if you do not, I will trample on you,'
" seemed visible in every word and feature.
" He appears to feel, in its fullest force, the
" sentiment,

> 'I have no brother, am like no brother:
> 'I am myself alone.' "

1065. It is unlucky for this blade, that the parties are *alive*. First—let the "*English wo-*"*man*" speak for herself, which she does, in these words:

1066. I remember, that, about a week after I came to Hyde Park, in 1817, a man came to the house in the evening, when Mr. Cobbett was out, and that he came again the next morning. I never knew, or asked, what countryman he was. He came to the back door. I first gave him a chair in a back-room; but, as he was a slippery-looking young man, and as it was growing late, my husband thought it was best to bring him down into the kitchen, where he staid till he went away. I had no talk with him. I could not know what condition Mr. Cobbett found the house in, for I did not come here 'till the middle of August. I never heard whether the gentleman that lived here before Mr. Cobbett, was an American, or not. I never in my life said a word against the people or the country: I am very glad I came to it; I am doing very well in it; and have found as good and kind friends amongst the Americans, as I ever had in all my life.

MARY ANN CHURCHER.

Hyde Park,
8th January, 1819.

1067. Mrs. Churcher puts me in mind, that I asked her what sort of a *looking* man it was, and that she said he looked like an *Exciseman,* and that Churcher exclaimed: "Why, you "fool, they don't have any Excisemen and such "fellows here!"—I never was at a *county court* in America in my life. I was out *shooting.* As to the *house,* it is a better one than he ever entered, except as a lodger or a servant, or to *carry home work.* The *path,* so far from being *trackless,* was as beaten as the highway.—The gentleman who lived here before me was an *Englishman,* whose name was *Crow.* But only think of *dirt, two feet deep,* in a kitchen! All is false.—The house was built by Judge Ludlow.

It is large, and very sound and commodious. The avenues of trees before it the most beautiful that I ever saw. The orchard, the fine shade and fine grass all about the house; the abundant garden, the beautiful turnip field; the whole a subject worthy of admiration; and not a single draw-back. A hearty, unostentatious welcome from me and my sons. A breakfast such, probably, as the fellow will never eat again.—I leave the public to guess, whether it be likely, that I should give a chap like this my *opinions* about *government* or *people!* Just as if I did not know *the people!* Just as if they were *new* to me! The man was not in the house *half an hour* in the morning. Judge, then, what he could know of my manners and character. He was a long time afterwards at New York. Would he not have been here a *second time,* if I had been familiar enough to relate *anecdotes* to him? Such blades are not backward in renewing their visits whenever they get but a little encouragement.—He, in another part of the extracts that I have seen, complains of the *reserve* of the *American ladies.* No "*social intercourse,*" he says, between the *sexes.* That is to say, *he* could find none! I'll engage he could not; amongst the *whites,* at least. It is hardly possible for me to talk about the public affairs of England and not to talk of some of my own acts; but is it not monstrous to suppose, that I should *praise myself*, and show that I believed myself destined to be the *Atlas of the British nation*, in my conversation of a few minutes with an utter stranger, and that, too, a blade whom I took for a decent tailor, my son William for a shop-keeper's clerk, and Mrs. Churcher, with less charity, for a slippery young man, or, at best, for an Exciseman?—As I said before, such a man *can* know nothing of *the people* of America. He has *no channel* through which to *get at them*. And, indeed, *why* should he! Can he go into the families of people at home! Not he, indeed, beyond his own low circle. Why should he do it here, then? Did he think he was coming here to live at *free quarter?* The black woman's hut, indeed, he might force himself into with impunity; sixpence would insure him a reception there; but, it would be a shame, indeed, if *such a man* could be admitted to unreserved intercourse with *American ladies*. *Slippery* as he was, he could not slide into their good graces, and into the possession of their fathers' soul-subduing dollars; and so he is gone home to curse the "*nasty guessing Americans*."

WM. COBBETT.

INDEX TO PART I.